POLITICAL MATTER

Political Matter

Technoscience, Democracy, and Public Life

Bruce Braun and Sarah J. Whatmore, Editors

A version of chapter 2 was previously published as Jane Bennett, "The Force of Things," *Political Theory* 32, no. 3 (2004): 347–72. Copyright 2004 Sage Publications.

Chapter 3 was published as William Connolly, "Materiality, Experience, and Surveillance," in *New Materialisms,* ed. Diana Coole and Samantha Frost (Durham, N.C.: Duke University Press, 2010). Copyright William E. Connolly. All rights reserved. Reprinted by permission of the publisher.

Chapter 6 was published as Nigel Thrift, "Halos," in *Mobile Nation: Creating Methodologies for Mobile Platforms*, ed. Martha Ladly and Philip Beesley (Waterloo, Ont.: Riverside Architectural Press, 2008).

Published by the University of Minnesota Press
111 Third Avenue South, Suite 290
Minneapolis, MN 55401-2520
http://www.upress.umn.edu

Library of Congress Cataloging-in-Publication Data

Political matter : technoscience, democracy, and public life / Bruce Braun and Sarah J. Whatmore, editors.
p. cm.
Includes bibliographical references and index.
ISBN 978-0-8166-7088-8 (hc : alk. paper) — ISBN 978-0-8166-7089-5 (pb : alk. paper)
1. Political science. 2. Technology—Political aspects. 3. Materialism—Political aspects. I. Braun, Bruce, 1964– II. Whatmore, Sarah J.
JA80.P63 2010
320—dc22
2010019695

Printed in the United States of America on acid-free paper

20 19 18 17 16 15 10 9 8 7 6 5 4 3

Contents

Acknowledgments

The conversation between science, technology, and society scholars and political theorists staged here began life as a workshop convened at the University of Oxford in December 2006. The chapters in this volume are the products of a workshop format in which their earlier incarnations were generative intermediaries in conversations that centered not on author presentations but on commentaries and exchanges organized around paired papers through which new and sometimes difficult connections, associations, and contentions emerged. We are extremely grateful to our authors for submitting to having their work forced through this mangle and have been reassured by those who stuck with the long revision process that they found the device as productive as the other participants.

We also acknowledge the participants in the workshop whose various contributions helped to shape our conversations but do not appear in this collection: Karen Barad, Mark Brown, Gail Davies, Paul Giles, Chris Gosden, Beth Greenhough, Dan Hicks, Steve Hinchliffe, Lois McNay, Derek McCormack, Annemarie Mol, Valérie November, and Steve Woolgar. We would also like to acknowledge the financial support of Oxford University Centre for the Environment for hosting the workshop. We are grateful to Pamela Richardson at Oxford and Elizabeth Johnson at Minnesota for their assistance in preparing the manuscript for publication and to Sebastian Abrahamsson for his assistance in preparing and recording workshop proceedings. Finally, we thank Jason Wiedemann at the University of Minnesota Press for his support for this project.

The Stuff of Politics: An Introduction

BRUCE BRAUN AND
SARAH J. WHATMORE

"THAT'S THE STUFF OF POLITICS." It is a common phrase that often denotes nothing more than the shady deals and overheated rhetoric that we imagine constitutes political life in representational democracies, as intimated by Sir Ivor Jennings (1962) in the title of the third volume in his study of party politics in Britain. Inflected thus, it speaks of the pursuit of power and all the tawdry practices that go into attaining it. It suggests something ignoble, even insubstantial—that's *just* the stuff of politics—an emphasis that separates politics from apparently more important or honorable aspects of human existence. But what if the accent lay elsewhere? What if we took the "stuff" of politics seriously, not as a shorthand phrase for political activity but to signal instead the constitutive nature of material processes and entities in social and political life, the way that things of every imaginable kind—material objects, informed materials, bodies, machines, even media ecologies—help constitute the common worlds that we share and the dense fabric of relations with others in and through which we live? What happens to politics—indeed to the "political" as a category—if we begin to take this *stuff* seriously?

This volume brings together science studies scholars and political theorists in an effort to address these questions and thereby to draw the insights of science and technology more fully into political theory and to bring political theory to bear more consistently on our understanding of scientific practices and technological objects.[1] Our objective is to sketch out a more fully materialist theory of politics, one that allows a place for the force of things and opens new possibilities for imagining the relationship between scientific and political practices and orderings. The absence of such a theory, we argue, leaves us unable to make sense of the collectivities in which we live and to respond adequately to the technological ensembles that are folded through social and political life. Without these conceptual tools, the profusion of complex materials with and through which we live too often leaves us oscillating between fearful repudiation and glib celebration. Such swings get in the way of creatively exploring new corporeal capacities or reflecting seriously on how we might have been, or could be, different than we currently are.[2] In convening this volume we thus set out from a simple premise: that science studies and political theory had much to offer such a project as well as each other.

This may indeed be a propitious time for such an endeavor. As we explain later, recent developments in science studies and political theory have resulted in the fields' convergence around a number of common questions, including the question of the company assembled in the name of the common itself.[3] Concurrently, the potency of technological objects and more-than-human agents in the fabric of political association and social conduct has become increasingly evident, and their force has registered more widely in academic and public life. From cell phones to stem cells, stuff of all kinds increasingly makes us what we are. Indeed, as many of the contributors to this volume attest, this materialization of the political has brought into view an ontological alliance between interests in the material propensities, affordances, and affectivities of nonhuman phenomena and the amplification of embodied human activity. The matter of politics and the politics of matter have never seemed so thoroughly entwined.

Yet, while science studies and political theory have much to offer each other, it seemed to us that their points of common interest had been left mostly implicit rather than directly engaged or closely interrogated. Examples of missed opportunities abound. For example, leading writers in science, technology, and society (STS), such as Brian

Wynne (2007), have repeatedly critiqued the peculiar tendency for scientists, often encouraged by public policy agencies, to treat the public as a homogeneous, preconstituted collective that can be more effectively mobilized or enrolled into scientific agendas by means of improved pedagogic or deliberative methods of engagement. But such critiques have not led to the wide-ranging or sustained exploration of political theories needed to help unshackle the notion and practice of public engagement from the dominant Habermasian (1996) model of deliberative democracy, in which speech is the only (and perfectible) medium of politics.[4] Despite their sophisticated posthumanist accounts of agency in laboratory life, far too many STS scholars remain stubbornly attached to humanist understandings of agency in public life. For their part, many political theorists (particularly, but not exclusively, those with environmental credentials) have had little difficulty in ascribing importance to technology as a political imperative for contemporary democratic practice—as something that ominously threatens the polis and that thus demands to be controlled.[5] However, by casting technology outside political life, they have had little to say about everyday technological practices or the lively materiality of technological objects within the collectivities in which we live. From this perspective, science and technology remain objects *of* politics—something we talk about—rather than something that inheres in and precedes the collective (and discourse), and thus something that challenges how the category of the political is *itself* conceived and where and in what it is articulated.[6] In similar fashion, the distinction between *zoe* and *bios* found in recent discussions of biopolitics (e.g., Agamben 1998; Esposito 2008) too rarely takes into account a third term—*technē*—without which the "becoming political" of our biological existence can hardly be conceived. Divorced from the things that constitute human life as such, biopolitics instead comes to be cast in ahistorical and metaphysical terms, unable to account for the retinue of objects and technical knowledges that condition the vitality of bodies and avail them to political calculability.[7]

Though it is tempting to claim that there is thus a yawning gap between science studies and political theory that this volume would then promise to fill, our intentions are somewhat different; our endeavor takes its inspiration from that extraordinary exhibition-compendium *Making Things Public: Atmospheres of Democracy* (Latour and Weibel 2005, 15) and its eclectic exploration of how "objects . . . bind all of us in ways that map out a public space profoundly different from what is usually recognized under the label of 'political.'" This is to say that we are less interested in finding a middle ground, or incorporating insights from one field into another, than we are in harnessing the frictions generated by the thinking of leading scholars in STS and political theory, and philosophers working at their interface, to spark new ways of understanding political matters and the matter of politics.[8] The objective of this volume, then, is to be inventive rather than integrative, to multiply the ways in which political technologies and technological politics are consequential in terms of what moves us to interrogate, contest, and engage with the force of things—in short, to become political.

What, we hear (some of) you ask, would move two geographers to instigate such an enterprise, and how, if at all, does it bear our disciplinary mark? The most ready answer is geography's historical insistence on understanding human life in relation to its material environments, a hallmark of the discipline that has given it its distinctive place in the academy (Braun 2007). More than this, however, geographers have had a keen eye for the significance of nonhuman entities and energies in the spatiotemporal arrangements of human life in terms of such persistent, if variable, practices as identity and memory, mobility and territoriality, and the shaping of proximity and distance (e.g., Massey, Allen, and Sarre 1999; Thrift 1996). In other words, though it has not been immune to the intensification of intellectual divisions of labor between the social and natural sciences, geography has never been comfortable with such divisions—a discomfort that has nourished a sustained insistence on and interrogation of the "more-than-human" fabric of social life and geopolitics (Whatmore 2002).

In this, geography shares much with kindred disciplines forged in the image of the human sciences that had purchase in the nineteenth-century academy before the bifurcation between social and natural science gained ground. With their shared focus on material culture, landscape, and human–environment relations, anthropology and archaeology also have continued to insist on the importance of "things," whether in terms of the symbolic dimensions of materials or their operational aspects, and on the ontological divisions and malleable relations between objects and subjects in diverse human cosmologies (e.g., Vivieros de Castro 2004; Gosden and Larsen 2007; McLean 2009). As the wider humanities and social sciences have (re)discovered an interest in the force of things in human affairs, so these more established disciplinary literatures have addressed them with new vigor (e.g., Oguibe 2004; Miller 2008) as well as the occasional complaint that their sustained body of work on this topic has not always been fully acknowledged (see Strathern 1996). It should come as no surprise, then, that geography too, particularly the kinds of cultural geography in which we are both invested,[9] has proved readily conversant with key STS concepts—like actant–networks, assemblages, and intermediaries (e.g., Thrift and Whatmore 2004; Anderson and Braun 2008)—while contributing to wider posthumanist currents in the social sciences and humanities (see Castree and Nash 2004).[10]

A more immediate (and personal) answer to the question of how our disciplinary location as geographers shapes this enterprise is that the profile of scholars that we have drawn into conversation here reflects the ideas and literatures that have been most influential for our own working through of the intersections between the practices, concerns, and intellectual resources of STS and political theory. We have found that each field offers much to such an undertaking. Science studies scholars, for instance, have increasingly taken as their concern the public life of science and technology.[11] Through detailed empirical studies, often ethnographic in method, these scholars have explored the composition of social, biological, and technoscientific assemblages—in sites as diverse as laboratories, law, and media—and insisted that we understand these assemblages to constitute dynamic conditions within which new understandings of the human, citizenship, and politics emerge. In part, this has meant taking nonhumans—energies, artifacts, and technologies—into account in the analysis of how collectivities are assembled, understanding these less as passive objects or effects of human actions and more as active parties in the making of social collectivities and political associations. This scholarship has also raised important questions about the spaces of scientific practice, the position of scientists as "representatives" of nonhuman constituencies, and the role of "experts" in making public policy. Much of this work has carried a sense of urgency, as scientific knowledges and technological objects have become increasingly controversial in public life and as science, technology, and politics appear to be ever more tightly intertwined in the everyday experience and social governance of processes as varied as biotechnologies, digital communications, and intelligent environments. Yet, though citizenship, democracy, representation, and politics are constantly invoked in this literature, it is not always clear to what these terms refer, which traditions in political theory inform them, or where these traditions might need revision.

Since its inception, political theory has also concerned itself with the composition of collectivities, or political association, whether understood in terms of sovereigns and citizens, publics and parliaments, or communities and nations. In its own accounts, however, political theory is often said to begin with (and as) an active purification of human society from the material world, in which the idea of humankind's removal from a state of nature marks the threshold of civilization and the possibility of political order—an idea prosecuted with vigor through projects of empire (e.g., Locke 1690; de Vattel 1760). In such stories the polis may well be understood as a place of lively public debate and its future understood as radically open to the play of political forces, but it consists solely of humans among themselves. The idea that "things" might condition political life is seen to return us to a primitive state, attributing magical qualities to inanimate objects. Despite this, we believe that modern political theory provides many openings to imagine the *matter* of politics differently. At least since Machiavelli, Hobbes, and Spinoza, it has understood collectivities—nations, peoples, and the state, or the relation between sovereign and multitude—in decidedly materialist terms, as a question of their ongoing assemblage rather than as primarily theological or philosophical questions. Such a point is stressed by Antonio Negri (1991, 1999), for instance, who locates in Machiavelli's *virtu* and Spinoza's *multitudo* the material conditions and conjunctive syntheses that enable political action (see also Althusser

2006). From such a perspective, citizenship, publics, state institutions, constitutions, and democratic assemblies are understood to be contingent outcomes and are to be studied as such; their formation is immanent rather than preordained. We find inspiration in such work, yet it is precisely here where we feel more needs to be done, for despite placing emphasis on political community as an emergent effect, modern political theory often imagines such associational achievements to be brought about by means of a social contract that binds individuals to one another by the force of words alone (e.g., Hobbes, Rousseau). Even approaches that emphasize preindividual or transindividual fields that precede the individual (e.g., Balibar 1994, 1997) tend to imagine these in anthropocentric terms. The effect has been to cast anything nonhuman out of the political fold or to relegate it to the status of resources or tools, entering political theory only to the extent that it has *instrumental* value but not in terms of its *constitutive* powers.[12] Science studies, we believe, has much to say to political theory about the everyday technoscientific practices and nonhuman objects that are party to the assemblage of common worlds, even as it has had far less to say about key concerns of political theory: sovereign power, political legitimacy, democratic citizenship, and public life.

Set out this way, we asked contributors to entertain a number of common questions as a starting point for a *posthumanist* political theory:

- Of what are collectivities and collective actions made? At what sites, through what practices, and by which actors? How do material and technoscientific objects contribute to such associative events and to their transformation?
- How is the more-than-human company involved in the reassemblage of social and political life to be addressed in theory? How do we register the affectivity of nonhumans in political life? What effect might this have on how we understand the category of the political?
- How is that which becomes included or excluded from collectivities determined? What sorts of institutional forms and political practices might be imagined to bring science

and technology into democracy, itself a contested term? Conversely, is democracy something that must account for "things," or are things there from the beginning?
- How is technology part of the art of government? Conversely, how should we think about governing technology?
- What is the relation between technoscience and its publics? Does the traffic between them only move from the laboratory into public life, or are publics active in the making of science and technology? If so, how and with what consequences for the politics of knowledge?
- What theoretical and philosophical traditions best provide intellectual resources for thinking the composition of common worlds?

The essays that follow do not directly answer each of these questions; indeed, some of the essays challenge their terms. Each, however, explores a specific aspect of what it means to conceptualize political matters in such a way as to include the matter of politics; taken together, they begin to articulate a new conceptual vocabulary for a more vital, and, it is hoped, more relevant, materialist political theory. Our task in the remainder of this introduction is to introduce some of the key concepts, issues, and questions that are put to work in the essays that follow.

Originary Technicity: The Becoming–Being of the Human

To call attention to the stuff of politics immediately brings us to a series of questions. What is this stuff about which we speak? Why is it important to ask after this stuff now, at this historical juncture? Finally, why should we imagine that our understanding of the political as a category might change if we attend to these matters? Answers to the first question will emerge in the book's individual chapters, populated as they are by pills, metals, plastic bags, vital systems, halos, and long-life bulbs. Let us concern ourselves here only with the others. Why inquire into the stuff of politics now? How should we name this juncture in which the question of the materiality of politics comes into focus? The most immediate and perhaps most self-evident answer lies in the sheer density of things that suffuse and shape everyday life: from simple tools and foodstuffs to smart cars, transgenic mice, new media, and pharmaceuticals. In such a context it is perhaps no longer possible to imagine either the human as a living being or the collectivities in which we live apart from the more-than-human company that is now so self-evidently internal to what it means to be human and from which collectivities are made.

But is this merely an effect of modernization, whereby life comes to be ever more mediated by things and thus ever further distanced from its biological or intersubjective constitution? In other words, is the story to be told that of the ever-increasing *colonization* of lifeworlds by an alien technology that threatens our individual and social autonomy? And is the relation between politics and technoscience, then, one in which the former names a realm in which we seek to come to grips with the threat of the latter? Though it is tempting to think so, and to imagine that globalization has spread this apparently modern condition to far-flung corners of the globe, the contributors to this volume categorically reject the idea that we have entered a novel historical period in which nonhumans, by sheer dint of numbers, now need to be included within our accounts of social and political life or in which what it means to be human is suddenly under siege by a world of scientific and technical artifacts that have come to shape who we are and what we may become. This is because they question human autonomy and self-sufficiency from the outset. Indeed, one of the arguments shared by all contributors, and perhaps the most significant point of departure for this volume, is that technicity—whether understood in terms of language, equipment, or machine—is not merely a supplement to human life; rather, it is originary.

This claim merits further discussion. What does it mean to say that human life is marked by an originary technicity? For Adrian Mackenzie (2002), originary technicity is a quasi-concept that helps us see that the association of humans and technical artifacts is more than just an external linkage by which the one comes to be connected to the other. The adjective originary, he explains,

> is one way to describe something more unnerving and unlocatable than merely strapping on, implanting or even injecting gadgets into

> living bodies. By now, "originary" has become familiar shorthand for the deconstructive logic of the supplement. The logic of the supplement describes all those situations in which what was thought to be merely added on to something more primary turns out to be irreversibly and inextricably presupposed in the constitution of what it is said to be added on to. (7)

In other words, it is a mistake to posit humanity as somehow separate from and existing prior to the world of things; rather, as thinkers as diverse as Jacques Derrida, Bruno Latour, Donna Haraway, Gilbert Simondon, and Bernard Stiegler have explained, the human comes into being *with* this world.

Such a view necessarily challenges how we think about the stuff that we consider properly technological, such as tools or machines, and the bodies that such entities are thought to supplement. Our stories often script technical objects as merely extensions of a preexisting body—the hammer, for instance, as an extension of the hand, which exists with an already formed capacity to hammer—rather than as objects that actively give shape to bodies and their capacities. Is it correct to see technological entities merely as a toolkit employed by a preexisting body, regardless of whether that body is an individual or a collective? Does it make sense to posit a fully formed hand with innate capacities *apart* from the objects that in a sense shape such a hand or afford it its capabilities, even if these objects are such basic things as sticks and stones? Drawing on the work of Gilbert Simondon and Maurice Merleau-Ponty, Mark Hansen (2006) argues that body schemas—the operational capacities of the embodied organism—necessarily involve the body's coupling to an external environment, a coupling that has always been accomplished through technical operations. Indeed, it is only from such embodied actions and encounters, he suggests, that a body image—a representation of the body—is secreted. As he puts it, "the noetic, representational body image is a derivative of—an emanation from—a more primitive, prenoetic bodily activity" (51). The natural or originary body that we take to be the starting point for political life is thus neither self-contained nor self-identical; to borrow the language of Derrida (1994, 141), when it comes to the body, we must always account for the *différance* of the technical apparatus.

In his discussion of body schemas and body image, Hansen's insight is not simply that the cognitivism that dominates the social sciences is misplaced, along with the privilege it accords to vision; rather, it is that the human body comes into being as such only in relation *to* the world. There is no moment at which humanity comes to be contaminated by technical objects and practices—no fall into a world of things—because there can be no human without them. The history of the human animal—and indeed the history of culture—is thus necessarily the history of the stuff that is, from the beginning, part and parcel of human life. Our embodied relations with things are not something that comes to be "added to" human life. The human body and its capacities emerge as such in relation to a technicity that precedes and exceeds it: there is no body, no original body, no origin outside this relation; no thinking, no thought, no logos, without that which forces thought. As we use the phrase, then, *technicity* refers to an exteriority that is necessarily also an interiority, or what various authors have discussed in terms of transduction, the coupling of embodiment and technics by which humans and nature interpenetrate (see Simondon 1992; A. Mackenzie 2002).

Crucially, if originary technicity makes sense in these terms, it also makes sense to understand it in universal terms. We do not mean by this that the history of technicity can be told everywhere in the same way, nor do we advocate a historical determinism that sees technology as the hidden hand directing the course of all human history. This is not universal history in a Hegelian mode; rather, to insist on a universal history of technicity is to insist that nonhuman and technical objects are an irreducible part of all stories of the becoming-being of the human, both individually and collectively, and that this could not be otherwise. It is to insist that the genesis of the individual (and here we understand the individual to mean groups and not simply organisms) is necessarily also a technogenesis.

Here it may be helpful to circle back to a previous question. Though the proliferation and potency of nonhuman objects in social life today may indeed render questions concerning the stuff of politics more intelligible than previously, we feel it is important to underline that the present context in which we ask the questions that shape this volume is not solely a scientific, technological, or social one; rather, it is also one that we might still wish to call ideological. The continuing difficulties that scholars face when writing from the perspective of an originary technicity suggest that humanism retains an extraordinarily powerful hold on the imaginative resources and analytical practices with which human life is thought. Most obvious, perhaps, is the stubborn attachment of many scholars—liberal and radical alike—to a humanism that finds ever new ways of positing the nonhuman as "out there," as Cary Wolfe (2010) puts it, rather than "in here," at the very heart of human becoming, and to a liberalism that continues to posit intention and action as attributes of autonomous individuals, rather than locating individuals and their capacities in relation to a larger transindividual field that precedes the individuation of singular things (Simondon 1992; Esposito 2008).

This humanist inheritance is compounded by the additional problem that specific disciplines have taken up the challenge of the posthuman or, as we would prefer, the *more-than-human,* in dramatically different ways. If geographers, anthropologists, science studies scholars, and philosophers have explored the question with renewed intensity, other fields have done so with far less enthusiasm. Arguably, political theory and economics can be counted among the latter. To the extent that political theory continues to assume that politics is something that occurs between humans alone, and economics holds on to the idea that the economy functions solely through the interactions of rational human actors, questions of science and technology, and the nonhuman more generally, necessarily remain a lacuna in both and continue to hinder our ability to understand how it is that our heterogeneous worlds are composed.[13]

The Performances of Things

From the concepts of *originary technicity* and *technogenesis,* it follows that we must be willing to speak of the performances of things and not just the actions of humans. Here we both draw on and depart from recent work in anthropology, art history, geography, and philosophy that has sought to understand "evocative objects" (Turkle 2008) or "things that talk" (Daston 2004). While we certainly agree with Sherry Turkle that material culture carries emotions and ideas of "startling intensity"—or that we think with things and not just about them, or even that self-creation occurs in and through our intimate relations to objects—there remains a surprisingly passive nature to the way such things are conceived. Daston is somewhat better in this regard, approaching things in a way that goes beyond examining their cultural meanings or their roles in subject formation. For Daston, things are not merely instruments for recording or playing back the human voice; they "talk," by which she means that they at once enable and constrain meaning: the language of things "derives from certain properties of the things themselves, which suit the cultural purposes for which they are enlisted" (15). This is an important insight that challenges what Daston describes as a Manichean metaphysics that asks us to choose, on one hand, between the "brute intransigence" of matter that is "everywhere and always the same" and its accompanying "positivist historiography of facts" and, on the other hand, "the plasticity of meaning" bound to specific times and places and its corresponding "hermeneutical historiography of culture." In an effort to think beyond this dichotomy, Daston argues that things must be approached as "simultaneously material and meaningful" such that matter "constrains meaning and vice versa" (16).[14]

Though we agree with Daston's critique of positivist and constructivist approaches to material things, our point is a somewhat different one: things are not just simultaneously material and meaningful; they are also *eventful*.[15] This political point needs to be unpacked further, for it is not enough just to say that things are lively and potent rather than dead or inert; rather, we wish to underline that things—and especially technological artifacts—carry with them a margin of indeterminacy. Their technicity is such that they can be combined and deployed in relation to countless other elements, gestures, practices, and institutions. Far from deterministic, technological artifacts temporalize, opening us to a future that we cannot fully appropriate even as they render us subject to a past that is not of our making. We are thrown forward into a future that cannot be foreseen, a future that "comes from behind" and thus challenges technocratic faith in our ability to know and control what is to come (see Stiegler 1999; A. Mackenzie 2002; Wills 2008; Diprose 2002; see also chapter 8).[16] It is precisely for this reason that new technologies are simultaneously celebrated and feared, for they can carry with them at one and the same time the promise of a glorious future and the threat of a catastrophic end.

As geographers, we would of course stress that objects do not only temporalize but that they also spatialize. In other words, it is not only that the technicity of the human opens us to a future that we cannot fully appropriate, it also brings about new assemblages and generates new spatial relations that at once contribute to this charge of indeterminacy and shape what is actualized at any given moment. Temporality and spatiality are thus intricately interwoven (Massey 2005). This is easily illustrated by way of something as ubiquitous today as the cell phone. It is relatively simple to see that its introduction opened being to becoming in new and unexpected ways. At the moment of its introduction, who could have fully known its effects? But if the cell phone temporalized, it did so because it radically transformed the topologies of everyday life—where and when we are reachable, dramatically redefining the capacity to be "in touch," with unforeseen consequences and possibilities for everything from relationships and work to politics and community. Indeed, one of the reasons that technologies temporalize is precisely because they spatialize, reordering our relation to other technical elements and to each other, regardless of whether these objects are light and mobile, like cell phones, or massive and immobile, like hydroelectric dams (Mitchell 2002; Massey 2005).

Political Matter(s) and the Matter of the Political

It is this excess of technogenesis that brings us to some of the central concerns explored by contributors to this volume. On one hand, the excess of technogenesis challenges our conception of the political as a category. If technological objects are objects that temporalize, then it hardly makes sense to locate them outside the political, if by this category we mean the practices by which our political associations and social life are constituted. Technologies are not just objects of political deliberation; they add their own dynamics to the differential relations that constitute social and political life. Nor can they be reduced to things on which decisions are made in the political realm because they are part and parcel of that realm from the outset. And they cannot be understood in terms of a future that is transparent and predictable because they are part of the reason that such futures cannot be known in advance. As many of the contributors to this volume show, the baroque nature of embodiment and technics, its folds and involutions, necessarily places us in the middle of things, without the certainties of humanism with its autonomous humanity or positivist science with its mechanistic matter. We are faced, then, with a confounding question: what, and where, is the "political" when emergent properties cannot be predicted, when all the actors cannot be known in advance, and when immanent causality necessarily bedevils political calculation?

Political theory has not been entirely silent on these questions. It is true that its contemporary paradigms have often privileged speech and human action or imagined that society is held together by a social contract that binds people together by force of words alone. We have already noted the problems associated with the deliberative democracy of Jurgen Habermas, which has been so influential on research efforts to operationalize deliberative modes of public engagement with science. But the problem is not limited to Habermas. The conceptual imperative of disagreement central to Jacques Ranciere's (1999) theory of democracy, for example, has been subject to related critiques (see Dillon 2003; Bennett 2005). Such understandings of political community have increasingly become frayed as political theorists have confronted the materiality of political life. We can see this already in the work of ecocritics like Thom Kuelhs (1996), who draws on the science of ecology and the philosophy of Gilles Deleuze to explode the bounded spatial imagination of the nation-state. For Kuelhs, our entanglements with biological life produce a complex topology that cannot be properly contained in an international system of nation-states. So, too, the philosopher Michel Serres (1995, 39) has addressed the consequences of the "exclusively social contracts" through which "we have abandoned the bonds that connect us to the world" and reworked the contractual polity toward an understanding of "the things of the world" and the "forces, bonds and interactions" in which they "speak" to us.

Of more immediate relevance has been the work of a number of scholars prominent in efforts to rethink the matter of politics across the social sciences and humanities and whose influence here is reflected in the chapters that form the opening section of this volume. These are the political theorists Jane Bennett (2001, 2005) and William E. Connolly (2001, 2007) and the philosopher Isabelle Stengers (1996, 1997). Each provides us with new conceptual vocabularies and thinking devices that, differently, articulate and extend the terms of a more fully materialist politics attuned to the "force of things" and capable of articulating those "things which force thought" in/as political practices through the convergent registers of affectivity, assemblage, and event. Bennett achieves this through a close and sustained engagement with the vitality of matter, particularly, but not exclusively, biological materials and bodily metabolisms, and by tracing out the ramifications of her "vibrant materialism" for political theory. Connolly does so through a focus on the affective power of things, particularly sensory media, to move us and shape our collective attachments and the political import of the ways in which this power is engineered and harnessed. From the very different trajectory of her demanding interrogation of scientific practices that succeed only insofar as the questions they raise are at risk of being redefined by the phenomena mobilized by an experiment or theory, Stengers extends this experimental ethos to elaborate an understanding of, even a test for, an adequate political theory (and practice) as one in which the things that force thought and attachment are active parties in political disputes.

Matters of Concern: Sparking Publics

Yet it is nothing other than the excess of technogenesis—its temporalizing dimension—that incessantly presses on us as a matter that cannot simply be cast aside, for it opens us to a future that is not known and cannot be anticipated. It is for this reason, rather than some sort of nostalgia for a non- or pretechnological humanity, that technological objects consistently emerge as "matters of concern" in public life. Latour's (2005) "matters of concern" refigure the political as an eventful technogenesis or, as he would doubtless prefer, *Dingpolitik*, amplifying the *res* of the *res publica* such that "the matters that matter in the *res* . . . create a public around it," triggering "new political occasions" (16). This returns us to Stengers's experimental "test" for scientific (and political) practice alluded to earlier, which demands that

> if we take seriously those nonhumans that are best characterized as forcing thought . . . what we need to think about and address is not the empty generality of humans as thinking beings but something we usually reserve for expertise, the correlate of the classical definition of political agency: humans as spokespersons claiming that it is not their free opinions that matter but what causes them to think and to object, humans who affirm that their freedom lies in their refusal to break this attachment. (chapter 1)

Even if these technological futures are not ones that we can fully appropriate, they are nevertheless ones that we cannot not enter into. How and why publics form as they do around such "matters of concern" is one of the questions that we seek to explore (Marres 2005; see also chapter 7). These include concerns for forms of human life, but crucially they also include concerns for what used to be considered the "outside" of human life—nature—but which is perhaps better thought of in a broader sense of geophysical and biochemical materials, entities and processes with which humankind and social lives are intertwined. This latter point is too often pushed to the side in posthumanist literature, which, somewhat ironically given its displacement of the "human" from the center of its ontology, often continues to proceed in terms of a decidedly humanist ethics, where it is human life and its possibilities that are taken to be paramount. Even work by materialists who seek to suspend the premises of liberal humanism—Negri, Balibar, and their collaborators, for example—still tends to give us a "multitude" that is decidedly humanist (as a collection of singularities whose individuation occurs in a "social" world). One of the goals of this volume is to begin to think through the terms of what Bruno Latour (2004a) calls "learning to be affected" and Donna Haraway (2008) speaks of as "response-ability," the kind of political and ethical thinking that is called forth by the capacity of all manner of things, human and nonhuman, organic and nonorganic, to affect and to be affected by others. To do so means understanding things of all sorts as forceful and to begin to consider what it means to represent this forcefulness in ethical deliberation and political practice.

Stengers's (2005) ethos of experimentation takes up this challenge by proposing a triple helix of representational moments in the practice and politics of mapping phenomena into knowledge. Here objectivity is a distributed (and by no means guaranteed) achievement that correlatively frames a phenomenon in ways that enable it to act as a reliable witness (object) to its experimental definition and produces experimental scientists as this phenomenon's reliable (objective) spokespersons. Each of these facets of representation relies in turn on a third, an experimental apparatus that exhibits a phenomenon in a reliable way and thereby both extends the potential company of witnesses and spokespersons and multiplies the trials to which their collective objectivity is subjected. Suspended thus in the weave of this delicate distribution of representational powers, the political charge of "things" becomes suggestive in a number of ways.

First are a set of questions about the experimental cast of the politics of knowledge and technology at work here. Its materialist insistence that the produced-ness of knowledge is something made—of stuff—as well as something made up or storied in particular ways introduces an interesting third party to the repertoire of representation. Both Stengers and Latour rely, as Lisa Disch (chapter 10) argues, on an intermediary device between the force of the world and the power of words—that of an indexical sign that transposes the force of a phenomena staged in an experimental event into a signal that does not gain its charge in relation to other signs but rather from the force it registers (e.g., a thermometer) and articulates. If such a material semiotics underpins the practices of experimentation, what are its implications for those scientific knowledge claims based on field- or model-based practices or, just as important, for the wealth of experiential or vernacular knowledge that informs the public reception and political contestation of the objects of science? Here we pick up a second set of questions concerning the political amplification of the force of things that centers on the relationship between what some-thing can do and what it can be shown to do, or how it can be made present again, beyond the experimental event. These issues have been addressed most directly by Noortje Marres (2005; also chapter 7), who, working with Dewey's (1927) notion of *material publics,* insists that the power of things to spark new publics into being and thereby to generate new political demands requires closer attention to the mediating devices through which they become affective. What are the registers and practices of public-ity—what Matthew Fuller (2006) calls *media ecologies*—that, in particular, technopolitical contexts amplify the power of things to move us and/or force thought among scientists and nonscientists alike? Film, plastic, computer games, and pharmaceuticals are among the answers that other authors in this volume explore (e.g., chapters 5, 6, and 8).

Finally, learning to be affected or to think response-ability also poses important challenges for scholarly thinking practices and is one of the creative tensions in the contributions to this book. In other words, the arguments that we have elaborated in terms of reworking political theory through an originary technicity imply a refiguring of political thinking through an ethos of practice. Though we would depart from Latour (2004b) in acknowledging that critique can be an inventive form of scholarly intervention, the onus of the contributions to this volume shares his insistence on the importance of multiplying styles and practices of scholarship into more diverse forms of political intervention and public engagement (see also Hinchliffe and Whatmore 2006). As we observed at the outset, critiques by STS scholars of prevailing "deliberative" modes of public engagement in the knowledge practices of (natural and social) scientists too rarely generate more inventive interrogations of political theories with which to experiment with other modes of public-ity. Contributions to this volume variously exemplify and/or signal a redirection of research energies and resources toward more constructive partnerships in the staging and practice of experimental knowledge polities in terms of the fora, media, and devices in and through which technoscientific objects are rendered affective and amenable to effective political interrogation. Some such experiments are under way in the guise of collective efforts to develop new research or pedagogic methods with which to interrogate, and intervene in, "knowledge controversies" of various kinds. One such involves an international collaboration between several universities[17] to operationalize Noortje Marres and Richard Rogers's (2005, 922) "recipe for tracing the fate of issues and their publics on the web." Another is a research project concerned with the knowledge practices and controversies associated with the science and politics of flood risk.[18] The project includes two trials of a method of collaborative research in which scientists (social and natural) and people who live with flooding and flood risk in the United Kingdom work together over a sustained period of time to generate new collective knowledge claims and competencies. Central to the working ethos and practice of this method (provisionally called *competency groups*) is "thinking with things," both in terms of interrogating the knowledge practice/technology of flood models as a key intermediary between flood science and politics and making a collective intervention in the public controversies associated with flood risk management in the areas in which the research is being conducted (Whatmore 2009).

What Comes Next?

The book is organized in three parts. The first sets out some directions of travel invited by prominent figures in the fields of philosophy of science and political theory whose work here, and elsewhere, is a reference point for many of the other contributions to this volume, including our own. This is followed by an arrangement of chapters designed to amplify a generative tension implicit in what we have discussed thus far between the register of technological politics in which the micropolitics, or minoritarian moments of civic contestation and innovation gathered in technological events, is the focus and the register of political technologies in which the macropolitics, or majoritarian moments of governmentality and political ordering, comes to the fore.

Part I, "Rematerializing Political Theory: Things Forcing Thought," contains essays by Isabelle Stengers, Jane Bennett, and William E. Connolly. Stengers, in chapter 1, opens the collection by arguing that including nonhumans in political theory is less a gesture of inclusiveness and more a recognition that nonhumans were never cast out of the political fold but rather that the very opposition between humans and nonhumans is itself witness to the idea of human exceptionalism to which political theory has been party. In other words, the problem of the inclusion of nonhumans in political theory carries with it the corollary demand to decenter political theory from the abstract concept of "humans," the consequences of which she goes on to explore. In the process, Stengers provides a rich grammar of minoritarian practices for refiguring knowledge controversies, like genetically modified (GM) technology, as experimental political events that gather the power that humans could not produce for "themselves" to achieve new collective thinking and invention. In chapter 2, Jane Bennett elaborates her concept of "thing-power" as a device to shift the political register of things, as congealments of matter–energy, from the epistemological terms of that which we cannot know to the ontological terms of agentic capacity or vital materiality. Making this journey with three companion-guides (a dead rat, a plastic cap, and a spool of thread), she harnesses diverse philosophical resources from Spinoza and Merleau-Ponty to Kafka and Margulis to work toward an event ontology that moves the project of a materialist political theory beyond the dialectical materialisms of Adorno and Althusser. Closing out part I in chapter 3, William E. Connolly works with the philosophical resources of Merleau-Ponty, Foucault, and Deleuze and Guattari toward an immanent materialism. Working through the consequences for a range of experiments and techniques that amplify the interinvolvements of sensory experience and perception, especially visibility, he elaborates an affective political modality largely ignored by political theory and practice.

These chapters raise a number of questions, which the chapters in subsequent parts of this volume pick up and take forward in theoretically and empirically diverse ways. How does their shared focus on originary technicity, the affective capacity and eventfulness of things, and their temporalizing political effects change our definition of the political? By opening up various possibilities for refiguring the political as the work of constitution or assemblage in which things force thought, association, and attachment, they provide rich resources for a posthumanist political theory that challenges the sway of the liberal subject as always already social or preceded by language by insisting that agency is distributed from the off. By the same token, they pose questions of their own, not least of which asks what is to be gained or lost through the vitalist tropes that writers like Bennett deploy. What, for example, do the metals, plastics, and halos with which authors work in part II of the volume suggest for the ways in which matter matters? Is more to be gained from a closer attention to the *specificity* of the matter at hand, as opposed to a generic analogy to "life" that could be described as a metaphysics?

Part II, "Technological Politics: Affective Objects and Events," is made up of three chapters that share, as intimated earlier, a common strategy of working their contributions to the questions raised by the "stuff of politics" through particular kinds of materiality. To this extent, they attend to the affectivities and affordances of specific, if mutable and indeterminate, things and the political events and practices to which these technologies give rise. In chapter 4, Andrew Barry explores the ways in which the informational enrichment of materials becomes a political matter. Using the example of metals, or more precisely, metallurgy, he interrogates the failure of a metal coating used in the construction of the Baku-Tiblili-Ceyhan oil pipeline as a political event through which parliaments and pressure groups are convened in a collective controversy across European borders. Working with the theoretical resources of Michel Foucault and Gabriel Tarde, he demonstrates how metals are far from docile objects and that metallurgists, as good materialists, expect the specific properties of metals to make a difference. However, he is equally insistent that there is nothing inherently political about them; rather, they acquire political agency only through their relations with others occasioned by particular contexts.

In the following chapter, Gay Hawkins turns our attention to that most ubiquitous of materials: plastic. More specifically, she traces three moments in the affective politics of plastic bags that afford very different kinds of political subjectivity and practice. Her goal is to interrogate how plastic bags come to matter without recourse to a materialist essentialism, arguing that their capacity to disturb and, in Connolly's terms, to compose new sensibilities is an associational achievement that complicates their political force and ethical practice. The final chapter in part II is by Nigel Thrift, who is concerned above all with the temporalizing effects of technicity as a necessary dimension of making room for the possibility of new political orderings. Using the conceit of the halo, he explores the multivalent political charge of things, complicit not only in the imaginative achievement of political orderings and associations but also in the articulation of alternative possible worlds, as well as to the powers with which both are resisted. Thrift works these concerns through three manifestations of the halo to exemplify the diversity of its affective grip on political imagining: the illuminated face of hallowed figures in Christian iconography, the interactive design spaces of the Halo computer game, and the occasional optical phenomenon of galactic halos.

All three chapters in part II place considerable emphasis on the capacity of things to interrupt political orderings and to open collectivities to new configurations. The four chapters making up part III of the book, titled "Political Technologies: Public (Dis)Orderings," share a different tack. The onus here is on the generative force of things in the assemblage and practices of political orderings and on the emergent publics and constituencies that exceed or challenge them. The chapters take up and interrogate the purchase of the "event" ontology entailed by political concepts like "matters of concern" and "situations" in terms of how this changes our grip on the techniques and forms of political association, mediation, and representation. Of particular importance here is a shared interest in the technological intermediaries that are party to the accomplishment of political (dis) order. In chapter 7, Noortje Marres derives conceptual resources from a close reading of John Dewey's formulation of the "public" to interrogate the import of "green technologies," such as long-life bulbs, and energy-efficient domestic appliances as political mediators, both as technologies of green governmentality and of inventive civic practices. She argues that Dewey identifies the rupture of everyday habits affected by intimate threats of material harm as critical events in the formation of publics that do not map onto existing social groupings. Moreover, for emergent publics to take hold, the material effects that call them into being require public-ity devices and techniques, such as electricity meters or energy standards, to render them widely observable and hence associative intermediaries.

In the following chapter, Rosalyn Diprose combines the resources of Foucault and Agamben's formulations of biopolitics with deconstructive phenomenology's concern with temporality to interrogate the synthetic steroid mifepristone (RU486) as a politically charged reproductive technology in the Australian body politic. She cites its capacity for political innovation in the tensions between its disciplinary effects in efforts to govern the reproductive (female) body and its temporalizing effects, refiguring the relationship between *zoe* and *bios* by opening up the disjunction between gestation (which involves a lapse of time) and birthing (giving time). Her chapter provides an important feminist interrogation of the relationship between *zoe, bios,* and *technē* such that the category of the political has to be reconceived through the differentiation of its body-subjects. The third chapter in part III is by Andrew Lakoff and Stephen J. Collier, who turn our attention to biotechnologies and to the intricate alliance between techniques of publicity and political technologies that articulate more obviously than most the reach of the state: infrastructure. More specifically, they are interested in the ways in which so-called vital systems (water, energy, transport, communications, etc.) foster new forms of vulnerability and have become objects of knowledge for security experts and civil defense agencies in the United States, a phenomenon they describe as a "political technology of preparedness." This is a political technology characterized not by the calculation of probabilities but by the imaginative enactment of events effected through various visualization techniques and practices of "mapping vulnerability." In the final chapter in the volume, Lisa Disch performs an original and exacting reading of the work of Bruno Latour and Isabelle Stengers and their efforts to remedy the dire representational consequences of the modern cleavage between science and politics and the settlement that this sets in train between the world and the word, mute things and speaking humans, facts and values. Her interrogation centers on the ways in which they set about redistributing the capacity for "speech" and the license to "speak for," arguing provocatively that they take their conception of representation from the assembly (the political spokesperson) but their practice from the laboratory (the experimental staging). Though she exposes flaws in the transposition of this "experimental" mode of representation to the political arena, she concludes that it should be possible for the represented to have a dialogic relationship with those who claim to represent them in both science and democracy based on openness to risk.

The final chapters return us to a generative tension that runs throughout the volume between two registers of technological politics and two ways in which the force of technological events in political life might be registered. For some contributors, technological politics are notable precisely because they signal an excess to political order and gather within them the possibility for minoritarian movements of civic contestation and innovation. For others, technological politics are part and parcel of majoritarian movements in which political subjectivities and orderings are actively constituted. What all chapters share, however, is a refusal to imagine technology as something that encroaches on political life from the outside. From their varied perspectives, the life of the polis must never again be thought of merely as humans gathered together without their myriad attachments. The "stuff" of politics is there, from the beginning.

Notes

1. The essays appearing in this volume were first presented at the Stuff of Politics workshop at the University of Oxford in December 2006.
2. This is a consistent refrain in the work of Adrian Mackenzie (2002) and Bernard Stiegler (1999).
3. Here and in what follows, we understand the *common* or *common world* as a contingent unification of bodies and things that provides the basis for various forms of life, both human and nonhuman. For writers like Antonio Negri (2003), the common is always beyond measure, always in a progressive process of formation that exceeds its capture in any fixed or final form. For Bruno Latour (2004c), the common world is a pluriverse rather than a universe, comprising a multiplicity whose assemblage is always provisional. Defined thus, the common cannot be reduced to the actions of humans alone.
4. Critiques span radical democrats like Laclau (2005) and Mouffe (2005), feminist political theorists like Benhabib (1996), and theorists of science and technology like Latour (2005) and Stengers (2005).
5. These include post-Foucauldian approaches to the politics of technology that acknowledge the capacity of objects to mediate political relations without attributing them political agency (e.g., Winner's [1986] "Do Artefacts Have Politics?") or that elaborate what Rose (1999) calls "technologies of citizenship" in relation to theories of environmental politics (e.g., Dobson and Bell 2006). For a critique of these, see chapter 7.

6 Thus, e.g., theorists of environmental democracy imagine a realm of democratic deliberation separate from the technological objects over which deliberation is said to occur. Indeed, the relation between technology and politics is oddly occluded in much environmental political theory (see, e.g., the collection by Dobson and Eckersley [2006]).

7 Here accounts of biopolitics might profit from the conjunctural analysis offered by writers such as Althusser (2006) and Read (2005).

8 A further dimension to the generative frictions that we sought to harness in the workshop were the differences in academic training and political culture between scholars in these traditions variously situated in Europe, North America, and Australia.

9 While interest in and contributions to the development of these ideas have been most intensive in cultural geography, they are not confined to cultural geography. For example, economic geographers have become interested in the ideas of actant–network, assemblage, and intermediaries through an engagement with the work of economic sociologists like Callon (1998) and Mackenzie (1996) and, more recently, political geographers like Featherstone (2007) and Routledge (2008).

10 There are multiple, diverging currents in this posthumanist turn in the humanities and social sciences (see, e.g., Halberstam and Livingston 1995; Hayles 1999; Wolfe 2003; Badmington 2004).

11 Outstanding examples in a large literature include Mackenzie's (1996) *Knowing Machines,* Barry's (2001) *Political Machines,* and Mitchell's (2002) *Rule of Experts.*

12 The work of Louis Althusser, and especially his discussion of "material ideological apparatuses," stands as an important early exception (see Althusser 1971, 1996). Recent writers who have gone beyond this include William Connolly (2001), Jane Bennett (2001), and Timothy Mitchell (2002). See also the work of the geographer Vinay Gidwani (2008).

13 The work of Karl Marx stands as an obvious exception. There is no absence of *things* in Marx's vast writings; inorganic nature, after all, is viewed as man's external body, and for Marx, society appears, at least on the surface, as a vast collection of commodities. But though Marx provides countless openings for a more robust materialism, things are not "eventful" for Marx; rather, they embody human

desires, intentions, and actions—they exist as "fetishes," as the expression of the inner logics of capital (i.e., the machines that replace human labor but that cannot themselves create value) or, finally, as the waste produced through capitalism's uneven development.

14 Bruno Latour (1999) makes a similar point in his essay "Circulating Reference" about the series of translations between "material" and "meaningful" involved in the mobilization of soil in the knowledge practices of soil scientists.

15 It may not be merely an accident, then, that objects and things are confused in her title. For Daston (2004), things are primarily objects.

16 Derrida speaks of this in terms of an "infinite responsibility" located between the "inheritance" and the "to-come" of a present that can never merely be itself.

17 See http://www.demoscience.org/.

18 See http://knowledge-controversies.ouce.ox.ac.uk/.

References

Agamben, Georgio. 1998. *Homo Sacer: Sovereign Power and Bare Life.* Stanford, Calif.: Stanford University Press.

Althusser, Louis. 1971. "Ideology and Ideological State Apparatuses." In *Lenin and Philosophy and Other Essays,* trans. Ben Brewster, 127–88. New York: Monthly Review Press.

———. 2006. "The Underground Current of the Materialism of the Encounter." In *Philosophy of the Encounter: Later Writings 1978–1987,* ed. Francois Matheron and Oliver Corpet, trans. G. M. Goshgarian, 163–207. London: Verso.

Anderson, Kay, and Bruce Braun, eds. 2008. *Environments: Critical Essays in Human Geography.* Aldershot, U.K.: Ashgate.

Badmington, Neil. 2004. *Alien Chic: Posthumanism and the Other Within.* London: Routledge.

Balibar, Etienne. 1994. *Masses, Classes, Ideas: Studies on Politics and Philosophy before and after Marx.* New York: Routledge.

———. 1997. "Spinoza: From Individuality to Transindividuality." *Mededelingen vanwege het Spinozahuis* 71: 3–36 (Delft, Netherlands: Eburon).

Barry, Andrew. 2001. *Political Machines: Governing a Technological Society.* London: Athlone Press.

Benhabib, Seyla. 1996. *Democracy and Difference.* Princeton, N.J.: Princeton University Press.

Bennett, Jane. 2001. *The Enchantment of Modern Life: Attachments, Crossings, and Ethics.* Princeton, N.J.: Princeton University Press.

———. 2005. "In Parliament with Things." In *Radical Democracy: Politics between Abundance and Lack,* ed. Lars Tønder and Lasse Thomassen, 133–48. Manchester, U.K.: Manchester University Press.

Braun, Bruce. 2007. "Theorizing the Nature–Culture Divide." In *Handbook of Political Geography,* ed. K. Cox, M. Low, and J. Robinson, 189–204. London: Sage.

Callon, Michel, ed. 1998. *The Laws of Markets.* Oxford: Basil Blackwell.

Castree, Noel, and Catherine Nash, eds. 2004. "Themed Essays and Responses on Posthumanism: An Exchange." *Environment and Planning A* 36, no. 9: 1341–63.

Connolly, William. 2001. *Neuropolitics: Thinking, Culture, Speed.* Minneapolis: University of Minnesota Press.

Connolly, William. 2007. *Democracy, Pluralism, and Political Theory.* London: Routledge.

Daston, Lorraine. 2004. *Things That Talk: Object Lessons from Art and Science.* London: Zone Books.

Derrida, Jacques. 1994. *Specters of Marx: The State of the Debt, the Work of Mourning, and the New International.* Trans. Peggy Kamuf. New York: Routledge.

de Vattel, Emmerich. 1760. *The Law of Nations or the Principles of Natural Law.* Repr., Washington, D.C.: Carnegie Institute, 1916.

Dewey, John. 1927. *The Public and Its Problems.* Athens: Ohio University Press.

Dillon, Michael. 2003. "A Passion for the (Im)Possible: Jacques Ranciere, Equality, Pedagogy and the Messianic." *European Journal of Political Theory* 4, no. 4: 429–52.

Diprose, Rosalyn. 2002. *Corporeal Generosity.* Albany: State University of New York Press.

Dobson, Andrew, and Derek Bell, eds. 2006. *Environmental Citizenship.* Cambridge, Mass.: MIT Press.

Dobson, Andrew, and Robyn Eckersley, eds. 2006. *Political Theory and the Ecological Challenge.* Cambridge: Cambridge University Press.

Esposito, Roberto. 2008. *Bios: Biopolitics and Philosophy.* Minneapolis: University of Minnesota Press.

Featherstone, David. 2007. "Skills for Heterogeneous Associations: The Whiteboys, Collective Experimentation and Subaltern Political Ecologies." *Environment and Planning D* 25: 284–306.

Fuller, Matthew. 2006. *Media Ecologies: Materialist Energies in Art and Technoculture.* Cambridge, Mass.: MIT Press.

Gidwani, Vinay. 2008. *Capital Interrupted: Agrarian Development and the Politics of Work in India.* Minneapolis: University of Minnesota Press.

Gosden, Chris, and Frances Larsen. 2007. *Knowing Things: Exploring the Collection at the Pitt Rivers Museum 1884–1945.* Oxford: Oxford University Press.

Habermas, Jurgen. 1996. "Three Normative Models of Democracy." In *Democracy and Difference,* ed. S. Benhabib, 21–30. New York: Columbia University Press.

Halberstam, Judith, and Ira Livingston, eds. 1995. *Posthuman Bodies.* Bloomington: Indiana University Press.

Hansen, Mark. 2006. *Bodies in Code: Interfaces with New Media.* London: Routledge.

Haraway, Donna. 2008. *When Species Meet.* Minneapolis: University of Minnesota Press.

Hayles, N. Katherine. 1999. *How We Became Posthuman.* Chicago, Ill.: University of Chicago Press.

Hinchliffe, Steve, and Sarah J. Whatmore. 2006. "Living Cities: Towards a Politics of Conviviality." *Science and Culture* 15, no. 2: 123–38.

Jennings, Sir Ivor. 1962. *The Stuff of Politics.* Party Politics 3. Cambridge: Cambridge University Press.

Kuelhs, Thom. 1996. *Beyond Sovereign Territory: The Space of Ecopolitics.* Minneapolis: University of Minnesota Press.

Laclau, Ernesto. 2005. *On Populist Reason.* London: Verso.

Latour, Bruno. 1999. "Circulating Reference." In *Pandora's Hope: Essays on the Reality of Science Studies,* 24–79. Cambridge, Mass.: Harvard University Press.

———. 2004a. "How to Talk about the Body? The Normative Dimension of Science Studies." *Body and Society* 10: 205–29.

———. 2004b. "Why Has Critique Run Out of Steam? From Matters of Fact to Matters of Concern." *Critical Enquiry* 30, no. 2: 225–48.

———. 2004c. *Politics of Nature: How to Bring the Sciences into Democracy.* Cambridge, Mass.: Harvard University Press.

———. 2005. "From Real Politik to Dingpolitik: or How to Make Things

Public." Introduction to *Making Things Public*, ed. Bruno Latour and Peter Weibel, 14–41. Cambridge, Mass.: MIT Press.

Latour, Bruno, and Peter Weibel, eds. 2005. *Making Things Public: Atmospheres of Democracy*. Cambridge, Mass.: MIT Press.

Locke, John. 1690. *Two Treatises on Government*. Repr., Cambridge: Cambridge University Press, 1988.

Mackenzie, Adrian. 2002. *Transductions: Bodies and Machines at Speed*. London: Continuum.

Mackenzie, Donald. 1996. *Knowing Machines: Essays on Technical Change*. Cambridge, Mass.: MIT Press.

Marres, Noortje. 2005. "Issues Spark a Public into Being." In *Making Things Public*, ed. Bruno Latour and Peter Weibel, 208–17. Cambridge, Mass.: MIT Press.

Marres, Noortje, and Richard Rogers. 2005. "Recipe for Tracing Issues and Their Publics on the Web." In *Making Things Public*, ed. Bruno Latour and Peter Weibel, 922–35. Cambridge, Mass.: MIT Press.

Massey, Doreen. 2005. *For Space*. London: Sage.

Massey, Doreen, John Allen, and Phil Sarre, eds. 1999. *Human Geography Today*. Cambridge: Polity Press.

McLean, Stuart. 2009. "Stories and Cosmogonies: Imagining Creativity beyond 'Nature' and 'Culture.'" *Cultural Anthropology* 24, no. 2: 213–45.

Miller, Daniel. 2008. *Materiality: Politics, History, and Culture*. London: Routledge.

Mitchell, Timothy. 2002. *Rule of Experts: Egypt, Techno-politics, Modernity*. Berkeley: University of California Press.

Mouffe, Chantal. 2005. *On the Political: Thinking in Action*. London: Routledge.

Negri, Antonio. 1991. *The Savage Anomaly*. Minneapolis: University of Minnesota Press.

———. 1999. *Insurgencies: Constituent Power and the Modern State*. Trans. M. Boscagli. Minneapolis: University of Minnesota Press.

———. 2003. "Kairos, Alma Venus, Multitudo." In *Time for Revolution*, 170–81. London: Continuum.

Oguibe, Olu. 2004. *The Culture Game*. Minneapolis: University of Minnesota Press.

Ranciere, Jean. 1999. *Dis-agreement: Politics and Philosophy*. Minneapolis: University of Minnesota Press.

Read, Jason. 2005. "The Althusser Effect: Philosophy, History and Temporality." *Borderlands* 4, no. 2. http://borderlands.net.au/vol4no2_2005/read_effect.htm.

Rose, Nikolas. 1999. *Powers of Freedom: Reframing Political Theory.* Cambridge: Cambridge University Press.

Routledge, Paul. 2008. "Acting in the Network: ANT and the Politics of Generating Associations." *Environment and Planning D* 26: 199–217.

Serres, Michel. 1995. *The Natural Contract.* Trans. E. MacArthur and W. Paulson. Ann Arbor: Michigan University Press.

Simondon, Gilbert. 1992. "The Genesis of the Individual." In *Incorporations*, ed. J. Crary and S. Kwinter, 296–319. Cambridge, Mass.: Zone Books.

Stengers, Isabelle. 1996. *Cosmopoliques.* Paris: La Decouverte.

———. 1997. *Power and Invention: Situating Science.* Minneapolis: University of Minnesota Press.

———. 2005. "The Cosmopolitical Proposal." In *Making Things Public*, ed. Bruno Latour and Peter Weibel, 994–1003. Cambridge, Mass.: MIT Press.

Stiegler, Bernard. 1999. *Technics and Time: The Fault of Epimetheus.* Trans. Richard Beardsworth and George Collins. Stanford, Calif.: Stanford University Press.

Strathern, Marilyn. 1996. "Cutting the Network." *Journal of the Royal Anthropological Society* 2: 517–35.

Thrift, Nigel. 1996. *Spatial Formations.* London: Sage.

Thrift, Nigel, and Sarah J. Whatmore, eds. 2004. *Cultural Geographies: Critical Concepts.* 2 vols. London: Edward Elgar.

Turkle, Sherry, ed. 2008. *Evocative Objects: Things We Think With.* Cambridge, Mass.: MIT Press.

Vivieros de Castro, Eduardo. 2004. "The Transformation of Objects into Subjects in Amerindian Ontologies." *Common Knowledge* 10, no. 3: 463–84.

Whatmore, Sarah J. 2002. *Hybrid Geographies: Natures, Cultures, Spaces.* London: Sage.

Wills, David. 2008. *Dorsality: Thinking Back through Technology and Politics.* Minneapolis: University of Minnesota Press.

Winner, Langdon. 1986. "Do Artefacts Have Politics?" In *The Whale and the Reactor: A Search for Limits in an Age of High Technology*, 19–39. Chicago: University of Chicago Press.

Wolfe, Cary. 2003. *Animal Rites: American Culture, the Discourse of*

Species, and Posthumanist Theory. Chicago: University of Chicago Press.

———. 2010. *What Is Posthumanism?* Minneapolis: University of Minnesota Press.

Wynne, Brian. 2007. "Public Participation in Science And Technology: Performing and Obscuring a Political–Conceptual Category Mistake." *East Asian Science, Technology and Society* 1, no. 1: 99–110.

PART I

Rematerializing Political Theory: Things Forcing Thought

1 Including Nonhumans in Political Theory: Opening Pandora's Box?

ISABELLE STENGERS

How to Define Nonhumans?

LET US START with the obvious problem—the impossibility of giving an adequate definition of the term *nonhumans.* I will present three obstacles that stand in the way of such a definition.

The first obstacle is that the negative, *non,* does not correspond to any unifying category because we cannot use any longer the category of object. Objects, as opposed to subjects, will necessarily lead us back to problems of knowledge, whereas we must deal with nonhumans as existents.

It is true that theories of knowledge and of existences were conflated when existence was derived from a divine creation, with the human mind created as the image of the creator God. However, with modern philosophy, the theory of knowledge has been redefined with the finite subject as its organizing center. As Whitehead (1968, 74) lamented, "the question—'what *do* we know?'—has been transformed into the question, 'what *can* we know?'" this last question being that of a censor, or a judge, whose first concern is to respect a divide between legitimate objects of knowledge and fanciful speculations. Among those fanciful speculations is the claim that one way or another we *do* know that nonhumans have an existence of their own, an existence that demands to be addressed and that may impose on us obligations and duties. Duties and obligations belong to the realm of what Kant called *practical reasons*, restricted to human subjects. Subjects, Kant stated, should never be used as means for our ends, the reciprocal, implicit statement being that we should never consider nonhumans as anything other than just such means.

However, as soon as we deny this grand divide, we are confronted with a disparate multitude. How to unify the Web; the AIDS virus; oil-devouring cars; hurricanes; neutrinos; the climate; genes; psychotropic drugs, be they legal or illegal; the great apes?

The second obstacle is that we may have to face the eventual demands of beings that were comfortably put away as creatures of human imagination. Pandora's box is open indeed. Beings that were excluded as speculative make their comeback, and we no longer have the appetite or the criteria of the censor to keep them at bay. Gods and goddesses, djinns and spirits are not objects for positive, factual knowledge; they do not even have the power to persuade all of us that they exist, in the way that Hurricane Katrina did—forcing even Bush to stop his vacations. However, to claim that the AIDS virus, neutrinos, or genes have such a power would put experimental sciences in the place of the old philosophical censor. When we do not deal with earthquakes, hurricanes, or tsunamis, which have the power to force unanimous recognition, only that which has acquired experimental recognition, that which has been able to satisfy experimental demands and to resist experimental tests, would be deemed truly to exist.

Accepting such a halfway solution would doom political theory. It would be forced to open itself to the question of living together with the creatures of technoscience that are put into action by those beings that have been accorded experimental existence. But it would leave outside the concerns of all humans, both individuals and populations, who *do* know that Gods, djinns, or the Virgin Mary matter. Those nonhumans would remain a matter of belief, protected by constitutional freedom to believe whatever you wish but asked to remain in the realm of private lives, with no public voice.

The third obstacle to defining the term *nonhumans* is that in the process we may well also lose the definition of the human as such. Indeed, if we admit the Virgin Mary, we may well have to admit also those existents we reduced previously to the status of human ideas. I am not speaking about the chimera we can deliberately forge. When a philosopher discusses the good criteria for excluding the unicorn, for instance, she is unlikely to confront anybody claiming that the unicorn exists; there is no spokesperson for the unicorn, and it is just an ingredient in the answer she is constructing to a philosophical problem. However, this same philosopher will claim that the problem demands a solution nonetheless. The unicorn is an indifferent case, but the problem has an existence of its own because it has the power to demand a solution. Such a power has for its best witnesses those mathematicians who attribute some sort of a Platonic existence to mathematical beings, with theorems expressing what they are able to force us to attribute to them.

If we take seriously those nonhumans that are best characterized as forcing thought rather than as products of thought, the idea that the mathematician at work is a human is not false, obviously, but it is a rather poor idea, deserving only a quick "yes, of course." What we need to think about and address is not the empty generality of humans as thinking beings but something we usually reserve for expertise, the correlate of the classical definition of political agency: humans as spokespersons claiming that it is not their free opinions that matter but what causes them to think and to object, humans who affirm that their freedom lies in their refusal to break this attachment, even in the name of some common good.

Let us indulge here in a quick connection with Karl Popper's third-world creatures. Popper's idea of three worlds may have been a bit simple, too directly bound into his primordial concern with disentangling what he called *objective knowledge* from the question of subjective beliefs, however well founded. But this concern has the power to induce a true philosophical jump, daring a connection between ontology and the pragmatic demand for relevance. If the questions that follow a definition of objective knowledge in terms of beliefs are not relevant, if they confine philosophers in a labyrinth of dreary paradoxes while the producers of such knowledge seem free of such difficulties, no compromise should be accepted. We do not "believe in" the truth of a theorem; the theorem has the power to have mathematicians accepting it and working with it.

However, well before Popper, Alfred North Whitehead led us back, on a similar topic, to a rather original version of Plato. Plato, he claimed in *Adventures of Ideas,* proposed a definition of humans on which every philosopher has implicitly agreed, whatever her claims, when engaging a philosophical practice. Humans would be those whose souls are moved by the erotic power of Ideas, a power to be distinguished from coercive force. Even those contemporary so-called naturalist philosophers, who ask us to accept that thought can be reduced to blind neuronal configurations, are witnesses for this power of Ideas because they trust that they can convince those they address, whose neuronal configurations are apparently sensitive to their arguments.

Plato's ideas are, as we know, part of an antidemocratic move, claiming that ideas are not ours to create and freely discuss. They rather authorize a measure of value by proximity and faithfulness, with philosophers who contemplate them at the head of the city, guiding those who live in the famous cave where they have only access to distorted reflections producing confusion and discord. However, we have also to recall that those who were allowed to take part in political debates in the Athenian democratic city were never humans as such, but citizens. Plato's proposition, even if it was meant to denounce democracy, is about humans and not citizens. In this, it can be understood as the ancestor of what would be accepted now as a consensual truth, that there is a defining feature that unites humans beyond our diverging cultures and opinions and that opposes us, as humans, to everything else. From this perspective, Whitehead's version of Plato's proposition has interesting, humorous consequences. What makes us human is not ours: it is the relation we are able to entertain with something that is not our creation. It should rather be said, following Whitehead's Plato, that those who now call themselves humans are thinking under the power of what can indeed be called an Idea, an Idea that causes them to define themselves as humans.

As Donna Haraway emphasized (in a joint seminar in Stanford, California, in April 2006), the sixth day of creation as told in Genesis 1:24–31 is also a story about human exceptionalism. During the same day God created not only Adam and Eve, in his own image, but also beasts of the earth according to their kinds and the cattle according to their kinds, and everything that creeps on the ground according to its kind. These creatures are defined not as individuals but according to a "kind" that prepares them for use and classification by Adam and Eve. The very definition of the creation act prepares and justifies the dominion given to humans over everything else on earth.

I think we are allowed to conclude from these three obstacles that including nonhumans in politics cannot be reduced to taking an explicit account of the role they would already play in the fabric of political association and public life. I would claim that nonhumans were never cast out of the political fold, because this political fold mobilized the very category of humans, and that this category is anything but neutral as it entails human exceptionalism at its crudest—reducing (against Plato and the biblical God) what causes humans to think and feel to human productions. From this standpoint, the very drastic opposition between humans and nonhumans would then itself be the witness of the unleashed power of this (nonhuman) Idea that made us humans, as it allowed us to claim exception, to affirm the most drastic cut between those beings who "have ideas" and everything else, from stones to apes.

I will now come back a second and last time to Karl Popper, who both related Plato to a totalitarian society and came to affirm the need for a pluralist ontology, with his third-world creatures transcending human opinions and convictions. The twist Popper gave to Plato's ideas is rather interesting, as the power of his third-world creatures is not that of a model warranting the possibility of human agreement without political debates but that of problems able to have humans going against their convictions and most plausible opinions. In both cases, mathematics was a privileged example, but the example plays two diverging roles. For Plato the mathematical demonstration exemplifies the ideal to be generalized, that of certainty forcing agreement, while for Popper it is the mathematical problem who (and the "who" here is important as it relates to the problem as an individual being) has the power to cause human thought to invent, for instance, irrational numbers or complex numbers. This corresponds to two different images of thought, as Gilles Deleuze would say, and this difference may be useful as I turn now to another aspect of the problem: Bruno Latour's proposition that we should treat humans as well as we are treating nonhumans.

Mistreating Humans

Such a proposition may seem a bit paradoxical if we remember cattle or the famous brevetted *oncomouse,* with whom Donna Haraway asks us to think: she who was fabricated to suffer cancers and be an object of biomedical experimentation for the eventual well-being of human women in rich countries. The paradox, however, gets clarified if we understand that Latour was addressing social scientists and that he was contrasting the way they deal with humans to the way experimental sciences deal with their nonhumans—molecules, electrons, or neutrinos. He would probably agree with my proposition to exclude from those experimental sciences most of biomedical research and to include among those who treat nonhumans well those scientists and nonscientists who engage in adventures of co-becoming together with apes, dogs, or crows. The best witness for such a co-becoming is again Donna Haraway, who now thinks together with her dog Cayenne, with whom she practices agility sport and because of whom, she claims, she learned more about herself, and about power, love, and ethics, than ever with human partners (Haraway 2008).

What is the common feature uniting let us say, for instance, an experimental physicist coming to the Nobel-prize winning conclusion that neutrinos have a mass, and Haraway experimenting with what it takes to become a partner with Cayenne and her achieving a performance that a (human) jury will assess? It may be that some social scientists would say "social recognition," and this is where the problem begins, because this would be an insult against both the physicists and Haraway—even if, as she says, Cayenne manifests pride and elation after a good run.

Social recognition is one of those blanket categories the use of which is the reason why Latour talks about social sciences mistreating humans. Obviously, when physicists got the Nobel Prize for their now massive neutrinos, this recognition was important for them, and we are familiar with the nasty quarrels that can be generated by authorship priority or due credit not being given to a colleague whose work had been an ingredient in a published argument. However, though physicists were not insulted by the Mertonian sociology of science, readily accepting the importance of social recognition, they would feel insulted if this were to become the social explanation for their involvement.

The outrage of scientists (remember the so-called Science Wars) is not exceptionalist or, at least, need not be. If a deconstructionist social scientist described a football game as no more than a matter of winning, with no difference between a beautiful collective move that results in scoring a goal and the corruption of players in the opposing team, he would be well advised to avoid sharing his "scientific" point with players and fans. Similarly, Donna Haraway tells us that those who judge or participate in agility sports will be pitiless in their opposition to those participants who would use punishing methods to train their dog.

Furthermore, in all these cases, the difference between the means is not, or not only, a matter of following rules if we associate rules with arbitrary connotations such that they could just as well have been otherwise. Nor is the difference a matter, or not only a matter, of obeying norms if we associate norms with ideally fixed values qualifying human behavior. Both rules and norms suppose that each particular case should be characterized in terms of conformity, whereas what I am trying to suggest is that the difference between the means is a matter of accomplishment—referring to an eventual *achievement.* Such an achievement does not designate a human being as such but as engaged with something else. When Donna Haraway learns how to address Cayenne, it is also and indissociably an achievement for Cayenne. As for the football player scoring a goal, he cannot be disentangled from an ever-changing relation with the other players and with a moving ball in a space put under tension by its limits, the goal line, and the goalmouth (Massumi 2002).

It is because an achievement is at stake that it is relevant to speak about the eventual reaction "we are insulted—this is war." To come back to what have been called the Science Wars, it did not surprise me. In fact, I had described it already as bound to happen in my *Invention des Sciences Modernes,* which was published in 1993. I knew that the sociological and cultural debunking of objectivity and its conclusion that objectivity was a matter of human agreement (with reality remaining mute, unable to make the imagined difference), whatever the experimenters claimed, was bound to be felt as a hostile attack. In itself such a claim was not something new—if you read Quine, you will find him concluding that as a philosopher, he is unable to define any intrinsic difference between the entities of physics and the gods of Homer. But Quine was defining himself as a philosopher who respects scientists, not as someone engaged in a demystification enterprise. In 1993 a derisive but rather cool Stephen Weinberg would write in his *Dreams of a Final Theory,* "To tell a physicist that the laws of nature are not explanations of natural phenomena is like telling a tiger stalking prey that all flesh is grass. The fact that we scientists do not know how to state in a way that philosophers would approve what it is that we are doing in searching for scientific explanations does not mean that we are not doing something worthwhile. We could use help from professional philosophers in understanding what it is that we are doing, but with or without their help we shall keep at it" (21–22).

My position, as I developed it in *The Invention of Modern Science* (Stengers 2000), is in complete agreement with this quotation from Weinberg, including his analogy with the stalking tiger. Rather than a question of norms, what we are dealing with here is the question of what makes both the tiger and the physicist "in their element" when stalking what they trust may be captured.

The irony of the problem is that the debunking urge and critical claim that a general statement about a matter of fact cannot be anything else than a human interpretation is as old as Galileo himself. It can be found in his *Discorsi,* at the beginning of the "third day discussion," when Galileo-Salviati will introduce the most famous definition of the "naturally accelerated motion." And it is voiced by Sagredo, who here as everywhere acts as a foil, at the service of Galileo's claims. Galileo needs this objection to promote the exceptionality of his definition of the motion of falling bodies. While all abstract definitions may well be arbitrary, produced by a human author, he will show that his definition is not: it is the definition of the natural motion of the falling bodies. In this case, and because of what I would call the first intervention (*intervenire,* "to come between") of an experimental device—the inclined plane—facts have been given the power to make a difference that no rival interpretation can explain away. From this point of view Galileo and the debunking critique are twins. Indeed, Galileo developed an exceptionalist rhetoric, proposing to put everything in the same bag as nondecidable human fiction, except that which can be decided through facts. The critiques just add to this same bag the "objective knowledge" that Galileo and his successors claimed to be the exception to the skeptical rule.

However, Weinberg's stalking image is not part of this rhetoric. His is a *cri du cœur* that situates him in the theoreticoexperimental history initiated by Galileo, when it was discovered that it may be possible to achieve a framing of natural phenomena that gives them the power to act as reliable witnesses for the way they should be defined. Just as the stalking tiger does not have a "neutral" definition of the prey it has in its sights and, eventually, in its clutches, so the experimental definition is obviously not the definition of the "phenomenon in itself"; rather, the experimental achievement (verified by colleagues) is to attribute to the phenomenon responsibility for its definition, correlatively producing the scientist as the phenomenon's reliable, "objective" spokesperson. Weinberg's *cri du cœur* does not express the confidence that this achievement is fated to happen, that experimental achievement is somehow a "right of reason." It expresses the trust that it can, and may, happen. I would claim that this is the trust that gathers physicists, just as it is the trust that there is fleshy prey to catch that is the very life and soul of the tiger.

However, the stalking image is not relevant for agility sports as practiced by Haraway and Cayenne, and it is dangerous for any of those scientific practices that must proceed out of the lab because, in the field, facts are important as eventual clues but are usually not decisive as such. Indeed, the naturalist's field achievement does not have the power of a definition in terms of well-determined variables. It is rather a narrative with no guarantee that it will keep its relevance in other apparently similar cases.[1] And it cannot enter into technoscientific tales in which reproducibility is conquered and measuring devices, industrial procedures, and everyday artifacts eventually follow from an active networking activity connecting experimental labs to their many outsides.

The stalking image, furthermore, is quite interesting because it exhibits the disaster that follows when the experimental achievement is turned into a normative method after the Galilean rhetoric—

a data-based method that would be a general definition of the scientific approach as opposed to mere subjective opinion. When Latour asks that social scientists treat humans as well as experimental scientists treat neutrinos or falling bodies, the point is not to treat them in the same way; quite the contrary. If neutrinos or falling bodies can entertain an analogy with the prey of the stalking tiger, it is because catching them is an event, an achievement: the experimental achievement demands its prey to be recalcitrant. "Catching" humans, for scientists, is not an achievement if they are too ready to answer the scientists' questions and accept the setting they propose. Furthermore, a true experimental setting enhances the abilities of nonhumans as actors (see, e.g., Latour 1999a) in a demonstration, while human scientists too often produce settings that play down this ability because their first ambition is to produce data that avoid the accusation of having been suggested by the setting. Humans as such lack recalcitrance, and any method that mimics experimentation is thus mistreatment.

But mimicking experimentation is not the only possible mistreatment. Physicists felt insulted by the blanket categories used by sociologists to critically interpret scientific agreement and objectivity, and they manifested public recalcitrance. This reaction can surely be avoided by "soft" qualitative methods that would never risk insulting anybody. But these methods also profit from the human lack of recalcitrance as they require confidence and goodwill. In contrast, Bruno Latour's proposition—that humans are to be treated as well as nonhumans—entails learning from this recalcitrance as it offers the chance to learn from the physicists how to treat them well. This does not mean at all treating them the way they ask to be treated but rather accepting this recalcitrance as a challenge. Accepting also that the lack of recalcitrance against sociological interpretation by other groups is a matter of serious concern, it may well be that sociologists have the rather dubious habit of taking advantage of a "scientists know better" submission. How else explain that they think it is normal for people to accept and try to answer any question they ask—even those that de facto define them as easy prey for insulting blanket categories?

My understanding of Bruno Latour's proposition thus reinforces my first remarks about the need, correlative to the problem of the inclusion of nonhumans, to decenter political theory from the abstract concept of "humans." As I already emphasized, humans have nothing to do together with nonhumans because their very definitions oppose them. But "human sciences," when they methodologically mimic the kind of objectivity that results from the event of an achievement in experimental sciences, demonstrate that the "denuded" humans required by their questions to provide access to "what is human in humans" are, above all, weak: complacently accepting to play the proposed role in the service of science. Are they not also playing a role, this time at the service of the functioning of democracy, when they fulfill their part through elections, for instance?

Diverging Minorities

What I call human Gilles Deleuze and Félix Guattari (1987), in *A Thousand Plateaus,* define as a "standard" (in French, an *étalon,* which means both a "standard" and a "stallion": meanings enjoined in the white, male, middle-class husband and father citizen). The standard hu-man has the power to define everybody else in terms of a deviation from what then becomes taken as normal. Yes, when submitted to the standard methods of social sciences, physicists were insulted, but it is because they are physicists. Yes, this one refused to answer my questions and even endeavored to criticize them, but it is because she belongs to that ecological group, and so on.

Marketing is already escaping such reasoning, using profiling techniques to target offers and information. However, what was interesting to Deleuze and Guattari was not the proliferation of microstandards but the positive, nonnumeric contrast between the majority and minorities.

If you accumulate specifications, whether targeting ten persons or one, you are still thinking in standard terms because you know that if you progressively suppress specifications, your group will become larger and larger: at the end comprising maybe the whole human species, with only deviant mavericks not joining the unanimous answer to questions such as "Would you choose to be rich and healthy rather than sick and poor?" In contrast, you are not free to define minorities and to ask them questions of your (methodological) choice: they define themselves together with the questions that matter for them. And they never dream that those definitions and questions are the "normal ones" but rather experience their irreducible entanglement in the process of their own becoming a minority group.

In short, the Deleuzian difference between majority and minority plays on the two distinct meanings of the term *ensemble* in French: "set," in the mathematical sense, and "togetherness." A mathematical set can be defined from the outside; all its members are interchangeable from the point of view of this definition and, as such, may be counted. But those who participate "together" in a minority group cannot be counted, as participating is not sharing a common feature but entering into a process of connections, each connection producing, and produced by, a becoming of its terms.

The examples Deleuze and Guattari use of minorities were all "dissidents," threats against public order. My use of the concept of minority may be seen as a betrayal, as I wish to downplay the original oppositional connotation and affirm its relevance for the togetherness of what I call "practices,"[2] whose members can be described as "attached" to something that none of them can appropriate or identify with—a nonhuman—but that causes them to think, feel, and hesitate.

Physicists as practitioners certainly do not present themselves as a threat to public order. They even present objective science as what separates society from the rule of might. However, like the Proustian Baron de Charlus, I would say to them, "You do not really care about society, do you—little scoundrels!" What they really care about, as physicists, is rather what Bruno Latour called *Knots* and *Links* (Latour 1999b): Knots achieved with what they address and interrogate, Links with colleagues whose role is to put the Knot to the test (to try to part what has been connected) but who will also produce consequences for the eventual achievement by connecting it with their own questions, making it an ingredient of new Knots. What Bruno Latour calls the Allies—those who have to be interested in scientific production for it to get both the means to proceed and the possibility to have consequences "outside"—know rather well that scientists are not very loyal partners, always prone to divert the resources they have obtained for aims of their own or to distract grounded interests in favor of speculative possibilities. As long as today's so-called new knowledge economy does not dissolve what links and gathers these scientists by putting each of them directly at the service of some partner's own priorities (as begins to be the case in biotechnology), the role scientists claim to play in the maintenance of public order will be a partial lie.

The concept of minority is relevant to affirm that any practice lies when it defines itself at the service of "human" needs or priorities. Correlatively, I would emphasize that if it indeed happens that minorities appear as a threat to public order, they are not defined by the desire to be such a threat, as would be the case with classical radical militant groups. What would rather define them is that they are threatened by the demands, rules, and priorities of public order.

I would define a practice through an attachment to a nonhuman in the enlarged sense (it could be a Popperian third-world creature, a river, or the Virgin Mary as the aim of a collective pilgrimage), an attachment that is not sentimental or habitual—I count on Thalys to go to Paris—but that has the power to make practitioners think, feel, and hesitate. Similarly, when Bruno Latour speaks of matters of concern, this does not just mean "to be concerned" by something. I may be concerned by a railway workers' strike if I intend to go to Paris, or by the AIDS epidemic, or by river pollution, but I know that my concern is about the same as anybody else's in a similar situation. As a philosopher, however, when I was working with the physicist Ilya Prigogine, our matters of practical concern were distinct. His was his struggle to produce a mathematical-physical formulation that would compel his colleagues to recognize that the arrow of time was to be accepted at the "fundamental level of physics." Mine was to become able to characterize the demanding, passionate character of his struggle without giving it a "metaphysical" character. You need to be a theoretical physicist for the arrow of time to be a problem, and as a philosopher, my problem was not this arrow. It was not to reduce the arrow of time to a "problem for physicists" either. It was rather to understand theoretical physics starting from the fact that this practice has given to the arrow of time the power to have its practitioners thinking and hesitating.

To develop the idea of practices, not in a descriptive way but as actively linked with the concept of minority, the importance of "hesitation" will be my starting point, and it will be connected with nonhumans through the concept of "obligations." This will lead to a thesis that will bring me back to a question of political theory: practices diverge, and their divergence, not to be confused with contradiction, makes them recalcitrant to any consensual definition of a common good that would assign them roles and turn them into functional parts of public order, whatever its claims to excellence.

Hesitation is what differentiates a practice from a normative or rule-following activity. This does not mean that practices are free from rules or norms; rather, in cases that matter, practitioners have to wonder if those rules or norms are not called into question because *there is something more important* than conformity. What is more important depends on the practice, but the concept of practice I introduce generically demands that nobody is able to set the rule, to appropriate the norm, and to a priori silence hesitation.

If hesitation gathers practitioners, it is because rules and norms are discursive expressions tentatively formulating something that has no definitive, authoritative formulation and hence does not communicate through obedience—which I call "obligations." Obligations communicate with the possibility of their betrayal. If ever a practice exhibited this possibility, it is that of the Quakers, who, as we know, did not quake in front of their God but in front of the menace of silencing what was asked of them in a particular situation, answering it in terms of preset beliefs and convictions. I was also impressed by the active and explicit "culture of hesitation" implied by the so-called consensus techniques experimented with by nonviolent activists in the United States. In all cases, the betrayal of obligations means that the situation has not gained the power to have those it gathers thinking, feeling, and wondering. Such a power is denied as soon as a situation is considered as covered by generalities authorizing recognition or as rule governed.

I have already used the formula "what causes them to think and feel," and I hope it was obvious that cause does not here relate to a cause–effect reasoning of the type "this caused that." What is rather implied is that as a rule, situations do not have the power to make us think and feel, and this was why Popper saw fit to affirm the irreducibility of his third-world beings to habits, convictions, conventions, and customs that all allow us to recognize how to deal with a situation without having us thinking or hesitating.

Situation is meant to be a neutral term, and it is now important not to use any word—such as the Popperian *problem,* for instance—that would privilege thinking as an intellectual business. In *La Vierge et le Neutrino* (Stengers 2006), I gave as an example of practice collective pilgrimages toward a place where the Virgin Mary is reported to manifest herself. This "situation" includes no theological discussions but rather pilgrims "preparing" themselves during the trip—that is, praying together, telling each other about their suffering and hope, and *learning to tell it* in the perspective of the eventual encounter with Mary. How to become able to receive and experience the grace associated with the presence and gaze of Mary is what the pilgrimage is about.

It is obvious that such an experience cannot be a reliable witness for the existence of Mary in the experimental sense as it is not meant to resist objections and counterpropositions. But in both cases my thesis is that we must resist the temptation to understand the achievement in terms of general categories. When "thinking" is related to a cause that forces thinking and feeling, it is not to be characterized in general terms, as a human production. Obligations express what a nonhuman, be it Mary or neutrinos, demands for a Knot to be created. About those Knots themselves, only one general statement can be made: those who enter into such a Knot, or trust that they may do so, will be insulted if a description reduces such a Knot to a human production, not allowing for nonhuman intervention.

Mary and the neutrinos entail divergent obligations but not contradictory ones as a contradiction implies a homogeneity of the terms, that is, in our case, a general, insulting reduction of all practices to human social activities, described by the same categories. More technically, if practices never contradict each other, it is because contradictions exist only between discursive formulations, whereas obligations may well be tentatively expressed through discursive formulations but they are not defined through such formulations. Mary and neutrinos, as "causes" (again, not as a cause that would define its effects and be defined by it!), refer to the possibility of a Knot that may eventually be achieved, that is, to an achievement that requires what humans "as such" have not the power to produce. Obligations communicate with the possibility of this achievement, with the trust that it may eventually happen. This is why obligations primordially entail hesitation—bearing on "how" one is obliged, and communicating with the possibility of their betrayal—and this is why, when obligations are concerned, no discursive formulation may be final: such a claim would suppress hesitation and give a purely "human" definition, or reason, to the achievement.

The "how" is a question that exposes, and puts at risk, those who are obliged. This also means that only those who are obliged may take the risk of experimenting with changes in the formulation of their obligations, because only they are exposed by the question of an eventual betrayal, leaving them soulless and lifeless as practitioners, like a captive tiger for whom the difference between live prey and chunks of meat would no longer matter.

My approach aims at activating the feeling that we live in a cemetery of already destroyed practices that have been unable to defend their obligations against the "outside," be it because of persecutory violence (remember witch hunting, for instance); soft pressure to conform to the demands of public rationality; deconstructive human sciences relaying in the name of science a consensual climate of derision; or direct capitalist redefinition in the name of progress (think of the so-called economy of knowledge and the already instituted techniques of assessing academic "quality"). *Vulnerability,* the definition of their environment as problematic and sometimes threatening, is a common feature of practices, and that which escapes this vulnerability, through defining itself against its environment, we call sects.

If practices answer to diverging obligations, their convergence as demanded by a politically defined "common good" is part of what threatens them. Scientific communities' traditional claims for autonomy can be contested, and legitimately so, as the claim obscures the dense connections with Allies and networking activities. Moreover, it is a claim that demands exceptional privileges, thus happily ratifying the destruction of other practices. And finally, it entails such niceties as the distinction between sciences that serve the technoscientific exploitation of the world (so-called applied research) and those that would serve humanity's true needs (so called blue-skies research). However, this claim must still be heard as it speaks to the vulnerability of practices; the way their environment may destroy them if its demands, whatever the intention, threaten to deny or dissolve their obligations. As such, practices put political theory at a bifurcation between two asymmetrical branches, a realist one and a speculative one.

Political Theory at the Crossroads

What I call the realist branch is realist because it ratifies what is the case anyway—that is, the process of destruction of what I call *practices.* It is even possible to define *technology* (a very difficult term to define) as requiring such destruction. Technology is not to be confused with interconnected technical practices. The *logos* is rather something that "happens" to practitioners, a networking operation that binds them and even commits them to the destruction of their own practice. This is obvious, for instance, with information and computing practitioners: what we call *technology* here characteristically demands of its practitioners that they work in ways that diminish themselves by producing procedures that make possible a chain of command to be faithfully executed. We can also think of genetically modified (GM) technology. Not only are genetically modified organisms (GMOs) not the crowning application of biology, they are the result of a rather poor molecular biology with inflated ambition (wait for the second generation!). Moreover, they herald the direct mobilization of biologists through the so-called economy of knowledge such that intellectual property rights are already a constitutive part of experimental research, including public research, with the power to stabilize the gene-oriented framing of biologists' questions. The point is not that scientific research would be in the process of being dissolved into engineering, because engineering itself is redefined in such a way that the difference between a technical achievement and a means for something else entirely is becoming radically undetermined.

In other words, the realist stance would be to conclude that all this business of practices and of nonhumans as what causes thought and feeling is part of the past and cannot have the power to force political theorists to think. It could be all the more plausible as the authority invested in "theory," when we speak of political theory, which may well rest on following already paved ways and ratifying the process of destruction of what does not recognize the supremacy of public language and majority reasoning. I would just emphasize that ratifying the process that destroys practices is also ratifying the impossibility of including nonhumans as this same process is depriving them of their spokespersons. From the standpoint of "technology" (as I have characterized it), there are no longer any spokespersons, only stakeholders with no obligations. We deal no longer with politics but with governance—with situations deprived of the power to force thinking as they are defined by stakeholders' vested interests.

The second branch I discern is speculative. Part of the question here is to turn the consequences of technology, which are usually reduced to a question of necessary adaptation, into a political problem. Although we are beginning to take account of the destruction of biodiversity and the loss of traditional materials and spiritual cultures, we have yet to learn how, politically, to address the process of the destruction of our own practices! Indeed, the technologies associated with ranking or productivity evaluation at all levels today, which are characterized by experts who (unlike referees in the traditional sense) are no longer meant to read and think, are in the process of blindly modernizing research practices, that is, of destroying research communities. As with all practices, academic ones were in a state of "just surviving." To emphasize this point is not to give a special importance to our own destruction. The speculative challenge concerns the difference that could be produced if scientific communities were to succeed, in this time of need, in addressing different allies than the traditional ones—the State and Industry—which have now betrayed this alliance. To succeed, a prerequisite is that they learn how to present themselves in nonmajority terms.

However, the speculative challenge also concerns the question of new forms of political activism that are part of what brings nonhumans into the political fray. New spokespersons are making their presence felt, and where technology is concerned, I for one would gladly present myself as a daughter to what I do not hesitate to call the *GMO event* in Europe. Naming what happened with GM as an event signals both an unpredictability (it failed to happen in the United States) and a capacity to transform present perspectives on both the past and the future. My standpoint would be that of a witness to what I learned because of this event, as having shared the collective transformative learning trajectory that made it an event.

The contestation of GMOs turned into an event when, far from isolating the mobilized opposition groups, the tentative answers of the public authorities served as both fuel and opportunity to extending the debate about expertise; the lack of reliable knowledge about consequences; the absence, or silencing, of field specialists; the limitations of the precautionary principle; and so on. As we know, this produced a reframing of GMOs themselves. Today they are no longer seen as worthwhile innovations whose risks should be accepted despite some possible problems. For many they are vehicles for intellectual property rights, the synonym of a capitalist expropriation strategy and of the unsustainable development that follows.

I would also present myself as a daughter of the possibilities associated with what I would call *minority techniques.* Those techniques are called *empowerment techniques,* but to resist the hijacking of empowerment by governance theory, it is important to emphasize that the stake here is emphatically *not* the empowerment of stakeholders but rather the empowerment of a situation: giving a situation that gathers the power to force those who are gathered to think and invent.

A situation, when defined in terms of the stable, vested interests of stakeholders, is always defined in majority terms, but when this situation gains the power to cause thinking, it induces a becoming that we may associate with the production of a minority—as none of the relations, knowledge, or agreements so generated can hold "in general" without this power. Stakeholders' gatherings can easily be assimilated by normalizing and normative procedures because they produce agreement and conventions that have the stability of the gathered interests. In contrast, empowering minority techniques are needed when this normalizing procedure is defined as a trap to be avoided because what matters then is a collective becoming that humans could not produce "by themselves" but only because of the situation that generated the power to make them think.

We are used to associating techniques addressed to humans with manipulation, implicated in the exploitation of human weaknesses and suggestibility. This is part of the human–nonhuman divide and of the correlative ideal that politics should be the affair of autonomous citizens, free of manipulation and gifted with personal convictions and ideas. I am personally very interested by these empowering techniques, and especially by those that dare to include nonhumans as a necessary ingredient of the collective becoming they experiment. For instance, when the U.S. neopagan witch Starhawk (see, e.g., Starhawk 1997) writes about the Goddess as an empowering presence, and about rituals as a matter of experimentation[3] by which to learn how to invoke and convoke Her in different situations, or about magic as a craft for transforming conscious awareness, the point is clearly not a matter of belief in some supernatural power—just as neutrino physicists do not need to believe that there exist massive neutrinos waiting to be characterized. In both cases, the point is an achievement that cannot be reduced to general, purely human, categories, an achievement which demands that humans do not feel themselves as masters of the situation, as responsible for what is achieved.

The challenge for political theorists may then be to learn how to situate themselves in relation to such "empowering" experimentations, the role and importance of which can easily be dismissed or assimilated into the background noise, unable to seriously disturb established power relationships or, equally, to be misrepresented as a model embodying a new ideal of democracy. What theorists have to learn is how to relate to something that involves true experimentation on its own ground—experimentation in the actual possibility of laypeople taking part in the construction of knowledge, questions, and choices about the future, that is, becoming able to appropriate or reclaim the role that is formally theirs in a democracy.

It should be taken for granted that the outcome of this kind of experimentation is uncertain and that the business of political theory is not to predict the outcome—prediction is about plausible developments, not speculative possibilities. What may be a matter of relevant concern for political theorists is the way activists' experimentations and actions are relayed; the irrelevant demands that may pervert them (do you have a program?); the irrelevant evaluations that promote them as "the voice of the civil society" or deride them as incapable of entering into a true confrontation with the real powers. The way theorists characterize such experimentations is part of this aspect of the question. There is no neutral position here. Critical thought that highlights the ways in which users' movements can be diverted into stakeholders' positions or tells disenchanted stories about the abandonment of local initiatives following the creation of procedures that required empowered participation has something redundant about it insofar as it relays and ratifies the general judgment that failure is the norm and that democracy is bound to remain formal. We face here the same challenge as the one imposed by what I called *practices*—approaching them from the point of view of their eventual particular achievement and what we can learn from it, not in the general terms a failure authorizes.

A Political Proposition

My proposition about political theory and the challenge it would face is not a neutral one. It is a way of relaying Gilles Deleuze's answer to the question of the distinction between right and left on the political scene. He gave this answer in 1985, at a time when the fashion was neither left nor right, and he resisted the fashion. For Deleuze the difference was not of degree, which would mean a common measure, it was a "difference in nature" *(différence de nature). La gauche,* he affirmed, vitally needs people to think, and this is not a problem at all for *la droite,* which rather needs people to accept situations, questions, and prospects as they are already framed (Deleuze 1997, 174). Indeed it is amusing that right-wing politicians loudly protest when it is recalled how their forerunners opposed now consensual laws and regulations, limiting, for instance, the exploitation of workers. In a way, their protests have some justification because of the difference in nature between the past situation, when workers were thinking and fighting, and the present one, when what they have conquered has been assimilated and incorporated into the normal state of affairs. As a result, Deleuze emphasized that *la droite* has a normal relation with state power and, when in power, will never be accused of betrayal, whereas *la gauche,* if it achieves political power, is by definition torn between the responsibility associated with the State and the proliferating questions and demands it should relay and take into account. This difference may be related to that between the majority and minorities, because when people are engaged in what Deleuze calls *thinking*—that is, not accepting the state of affairs in the (majority) proposed terms—it always means the emergence of diverging minorities.

Deleuze's proposition opposes any prospect corresponding to a reconciled society in which everybody will feel free to indulge in her own affairs and personal development under the benevolent care of the State, with some nice intersubjective debates between well-disposed citizens about priorities and values. Opposing such a prospect does not mean having insuperable conflicts become the dark truth that this nice order would obscure or repress. It is again a question of a difference of nature, this time between two images of collective thought (and for Deleuze and Guattari, any thought is always collective), the tree and the rhizome. The tree is the State ideal because it is the image of problems ordered from the more general to the more particular: a "leaf" question will never cause a "trunk" agreement to be contested. It is also the figure of discursive logic: if we agree on such and such, then only this or that remains open to discussion. In a rhizome, any two points may get connected without hierarchy, without the irreversible if *x,* then *y,* and nothing, no signification, can be characterized as settled. However, the rhizome is not a figure of sheer anarchy: for Deleuze and Guattari the brain itself functions as a rhizome, and there is no question but that it is functioning. If nothing is settled in a rhizome, if we have to follow and not deduce, it is because connections are not redundantly following an order that they merely make explicit: new connections may add dimensions and transform the very pragmatic identity of what gets connected. A connection is an event, never a derivation.

I would propose that including nonhumans in the guise I have characterized—that is, as causes for thinking—both leads to a rhizomatic situation and protects the rhizome image against any assimilation with a network, such as a technological one, when each connected term has for its only identity the way it is connected with others.

The production of rhizomatic connections must be characterized as events, and to do so, I will try to situate it in the frame of an *ecology of practices.* Using the term *ecology* means that practices are to be characterized in irreducibly etho-eco/logical terms—that is, in terms that do not dissociate the *ethos* of a practice and its *oikos,* not only the matter-of-fact environment but the way it defines its relation with other practices and the opportunities of the environment. From this point of view, new connections or a changing connection, or a change in the environment, are events indeed, a possible transformation of what we would have been tempted to accept as the identity of a practice. From this point of view, also, scientific practices, when they present themselves in majority terms, correspond to the simplest case of ecology—that of predator–prey. All scientific practices agree to define what is not scientific as prey, but each is potential prey for another that can claim to be more objective (chemistry has been defined as "applied physics only") and a predator for other, so-called weaker ones (neurosciences define in predatory terms the domain of human sciences). Connecting events, on the other hand, must not be characterized as a matter of goodwill but as achievements, the creation of new possibilities and new questions for the concerned parties. Finally, the idea of an ecology of practices entails that each practice has indeed its own recalcitrant, diverging manner of defining what matters, what I previously characterized in terms of obligation. The point is that there is no direct connection between such manners and the definition of a well-defined ethos. The ethos may be defined only in relation with its oikos. In other, Spinozist, terms, "we do not know what a practice can do." Such an idea is not a ready-made political program, for sure, because it addresses practices as if they were recognized in their guise of recalcitrant minorities. This is a very big "as if," indeed, which links ecology of practices with political fiction or speculation. However, as such, it may serve as a tool for thought, a tool to orientate thought, propose constraints, and help to resist some utopian dreams—in this case, dreams that would presuppose the taming of nonhumans and the freedom of humans to decide how to live together through intersubjective communication.

The idea of an ecology of practices openly refers to the wisdom of naturalists who have learned to think in the presence of ongoing indicators of the past destruction of living species and will never accept any justification of such losses as the (unhappily necessary) price or condition for the progress of life on this earth. In other words, the ecology of practices openly refuses Capitalist, Marxist, and commonsense judgments about "practices condemned by History."

However, the ecology of practices is not a naturalist idea but a speculative one, because the very term *practice* is not descriptive. It does not address practices "as they are"—physics as we know it, for instance—but as they may become in different surroundings, when the analogy with interacting living species would become relevant in two senses: first, that belonging to a species means having a particular standpoint on one's environment, which means, for instance, that practitioners would be derided if they were to claim the privilege of speaking in the name of a general, transcending cause (human progress, common good, reason, the laws of nature, etc.); second, that in the absence of any general relations, nothing produced in a particular practical setting can be attributed a meaning or a value that would logically or consensually impose itself outside this setting.

For the analogy not to be a naturalizing one, it must be emphasized that even in natural ecology, the identity of a species is only a first approximation of the ethos of those belonging to a species; what they need, how they relate with each other, and their environment have no biologically grounded specific definition. Belonging refers to constraints, and each new variant ethos creates a new meaning for these constraints. Because of the passionate work of experimental ethologists exploring, for example, what kinds of active environments great apes require to learn, we even have to include into the contemporary definition of what an ape ethos may be those relational habits researchers call "speaking." As for the agility sport practiced by Haraway and Cayenne, it attests to a long story of co-learning entailing new kinds of ethos on both parts.

An ecology of practices is speculative because the whole point of it is the difference between belonging and identity, or the difference between obligations and fixed rules or norms that I have linked to the questions nonhumans impose on their practitioners. Obligations are constraints; they entail that a new practical ethos, with transformed relations with the environment, is a creation, not a change, that would be a function of the environment. Such an ecology of practices takes as its motto a practical version of Spinoza's famous dictum about the body: we do not know what kind of relation their obligations make practitioners able or unable to entertain with their environment.

For the concerned practitioners, however, it is not a matter of knowing or not knowing. What is at stake is an event—an event that puts into question what Bruno Latour (2004), in his *Politics of Nature,* distinguishes as *essence* and *habit.* This distinction is never settled, and each time it is in question, settled habits, what makes the identity of practitioners, are bound to protest with the essential cry, "If you try and modify us, you kill us." I would propose that a crucial speculative point in the proposed ecology of practices is what might be called a *culture of hesitation* that enhances the distinction between obligations and rules or norms, or between essence and habits. The point would not be to denounce habits and rules but to define as important, as worth the positive attention of practitioners, the amplification of this distinction.

However, such an amplification does not imply a general reflexive stance. To be related to the possibility of an event, hesitation must be concrete, bearing on the possibility of new or modified connections of practitioners with their outside. This is why, for instance, the ecology of practices will never be concerned with a change in the relation between science and the public, two abstractions that are made to meet only through generalities (information, miscomprehension, goodwill, pedagogy). The point would rather be the relation of instituted, hardwired practices (sciences, law, medicine, etc.) with empowered minorities who have become collectively able to object, question, and impose as mattering aspects of situations that would otherwise be mistreated or neglected.

The idea of ecology entails another consequence: the refusal of any transcendent standpoint. No definition of the common good, no appeal to reason, and no ideal of peace can authorize an arbiter when conflicting demands clash. This, however, does not mean blind clash or insuperable contradiction. As I already emphasized, divergence is not contradiction. It just means that any agreement has the character of an event, which may well be an answer to a common matter of concern but without the concern having the power to define its eventual practical consequences. Correlatively, eventual agreements will always be local agreements between parts that keep diverging: a pact, not a convergence with a common aim overcoming the divergence. In other words, we deal here with rhizomatic connections, which cannot authorize a treelike representation with the trunk standing for what diverging practices would have recognized that they have in common—for instance, as human practices.

As Deleuze wrote, an idea always exists as engaged in a matter—that is, as "mattering" (we have an idea in music, or painting, or cinema, or philosophy, or . . .). A problem is always a practical problem, never a universal problem mattering for everybody. If the ecology of practices entails that we do not know what kind of relation their obligations make practitioners able or unable to entertain with their environment, it also entails learning about it. And learning here is always local because the rhizomatic connections practitioners are able or unable to forge do not obey general rules or reasons. I have named as diplomats those practitioners whose obligations designate the possibility of generating rhizomatic connections where conflict seems to prevail.

The Art of Diplomacy

The art of diplomacy, in contrast to what prevails today, presupposes that the affronted parties must be defined in terms of forces that all decide to "give peace a chance," that is, that all agree on a slowing down of all the good reasons everybody has to wage a justified war. This excludes negotiated surrender, the aim being in this case not to envisage the possibility of peace but rather to envisage the economy of costly military operations. It may well be that this art is no longer relevant for international relations, but it still relevant to introducing the challenge of an ecology of practices.

To speak about diplomacy is to speak about borders and the possibility of wars. Borders do not mean that connections are cut but that they are matters of arrangement. Reciprocity itself, if it exists, is part of an arrangement, with different risks and challenges for each involved party. Free circulation and general equivalency, that is, the disappearance of the need for diplomacy, mean devastation in the terms of an ecology of practices. But the imposition of the rules of public language for a protest to be taken into account also means such devastation as public language defines all speakers as ideally interchangeable.

As such, the art of diplomacy does not refer to goodwill, togetherness, the sharing of a common language, or an intersubjective understanding. Neither is it a matter of negotiation between flexible humans who should be ready to adapt as the situation changes. It is an art of artificial arrangements that do not exhibit a deeper truth than their very achievement—the event of an articulation between protagonists constrained by diverging attachments and obligations in situations where contradiction seems to rule, a rhizomatic event without a ground to justify it, or an ideal from which to deduce it.

Such events have nothing to do with heartfelt reconciliation; neither are they meant to produce mutual understanding. Indeed, they are such that each party may entertain its own version of the agreement, just as in the famous example given by Deleuze of the *noce contre nature* (marriage against nature) of the wasp and the orchid. We get no wasp–orchid unity, he emphasizes, as wasps and orchids give each other quite another meaning to the relation that takes place between them.

This is why the art of diplomacy is usually despised as an art of hypocrisy and artificiality. Heartfelt reconciliation is then glorified as the only true way to achieve peace, against artificial constructions and arrangements. The rejection of what is despised as artificial is part of a very old legacy. The Christian faith in the saving power of truth—come and you will be free—as it is turned into the war cry of missionaries battling against "fabricated idols" testifies to the power of truth as defined against artifacts. Diplomacy is much older than Christianity, and it celebrates another conception of truth, a fetishist one, to refer to Bruno Latour's (1999c) very important analysis of our inveterate antifetishism as one version of the Great Divide.

The condition for diplomacy—that affronted parties slow down—means that they accept the possibility that a diplomatic proposition may eventually result in an arrangement, articulating what was a contradiction leading to war. But acceptance of this possibility does not mean acceptance of the proposition issued from the diplomats' encounter. This is why a diplomat will never say to another diplomat belonging to the opposing party, "Why don't you just agree with this or that proposal" or "In your place I would . . ." Diplomats, if true to the art of diplomacy, know that they are all at risk and that they cannot share the other's risk. Will the kind of modification on which may depend the possibility of the peace they are negotiating be accepted by those people each diplomat represents? Or will the diplomats be denounced as traitors when they return home?

Thus another condition for the art of diplomacy is what I call a *culture of hesitation*, the capacity for the protagonists not to confuse belonging and identity, that is, not to take as a betrayal or a manifestation of weakness the acceptance of a proposition that implies a modification of their habitual formulation of who they are. As we know, radical direct democracy is often associated with the idea of an imperative mandate and the disavowal at any time of a representative who would betray it. This is an interesting and challenging proposal if, and only if, representatives can trust that those they represent will be interested in their account of the situation and know how to hesitate and consult before concluding that the mandate has been betrayed. If the notion of imperativeness excludes hesitation about the way the imperative is to be satisfied, the representative is a hostage, and the proposal is self-defeating.

Diplomats are not theorists, and they speculate little about the potential role of political theory for the practice of their art. However, it may be that the idea of an ecology of practices, and the correlative art of diplomacy, may serve as an active test for the very idea of a "political theory." The recalcitrant groups experimenting with what it takes not to be wiped out by the normative leveling of divergences may well be considered as producing something that should matter for political theorists, as the very nature of a political problem may be transformed if those who are concerned by this problem produce the ability to play an active part in the way it is formulated. But the active test is that political theorists must also accept that they themselves are part of the always problematic and often threatening environment of those experimentations. The art of diplomacy may be enlightening here because it is an art of divided loyalty, binding diplomats both with the group they represent and with other diplomats, whose loyalty is also divided. In the political theorists' case, the point is not to represent, or to become, spokespersons. Yet divided loyalties remain necessary to sustain both the possibility that is concretely risking its actualization in such groups and the matters of concern that make the theorists themselves think. Such a position for theorists is not paradoxical. In fact, it already characterizes the cooperative symbiosis that exists between theoretical and experimental practices in sciences like physics and chemistry. Symbiosis is always between heterogeneous beings: theorists and experimenters need each other, but their practice entertains diverging constraints and obligations, which means that an achievement, or an interesting failure, entails different consequences for them. Symbiosis is about a culture of divergence, avoiding, for instance, what I would call theoretical voyeurism, when theorists do not respect the necessary distance and propose interpretations, the production of which is part of the experimentation and not the business of the theorist.

Such a distance has nothing to do with a form of neutrality. It is produced by the divergence of practices. However, for such a symbiotic culture of divergence to exist, *both* parts have to diverge. The question, then, is what causes political theorists to think and diverge for their own sake and, by the same token, what makes them accountable and in relation to what? This is the question of the obligations proper to their practice, and it may well be that it is in learning the demands of symbiotic relations with recalcitrant groups that such questions, which become vital in this case, may find the empowering beginning of an answer.

Notes

This text is an attempt to take seriously one of the questions addressed to the participants of the Stuff of Politics workshop from which this volume arose: what challenges does the inclusion of nonhumans hold for democratic theory? This is why it takes for granted, right from the beginning, that such an inclusion may be a matter for serious thought—at the risk of sounding like speculative (political) science fiction.

1 See about this Gould (1990). Gould compares the eventual achievement of coherence by the puzzled naturalist facing many disparate eventual clues with the "integrative insight" of Lord Peter Wimsey and not with the logical, deductive reasoning of Sherlock Holmes.
2 I began working with this term in *Cosmopolitics* (Stengers 1996–97) and it is at the center of *La Vierge et le Neutrino* (Stengers 2006).
3 That the same word, *experimentation,* is used for experimental sciences and for those minority techniques highlights two proximity points. The first point is that it is again a matter of *Knots* and *Links:* of the reciprocal causality between Knots created with a nonhuman and Links created among those who face a particular situation. The second point is that in both cases, we do not deal with *beliefs,* or revelation of a hidden truth, but with a deliberate pragmatic of the artificial, the truth of which rests in the contract between achievement, or efficacy, and failure.

References

Deleuze, Gilles. 1997. *Negotiations 1972–1990.* New York: Columbia University Press.
Deleuze, Gilles, and Félix Guattari. 1987. *A Thousand Plateaus.* Trans. Brian Massumi. Minneapolis: University of Minnesota Press.
Gould, Stephen J. 1990. "Triumph of a Naturalist." In *An Urchin in the Storm*, 157–68. New York: Penguin Books.
Haraway, Donna. 2008. *When Species Meet.* Minneapolis: University of Minnesota Press.
Latour, Bruno. 1999a. "From Fabrication to Reality." In *Pandora's Hope*, 113–44. Cambridge, Mass.: Harvard University Press.
———. 1999b. "Science's Blood Flow." In *Pandora's Hope*, 80–112. Cambridge, Mass.: Harvard University Press.

———. 1999c. "The Slight Surprise of Action." In *Pandora's Hope*, 266–92. Cambridge, Mass.: Harvard University Press.
———. 2004. *Politics of Nature: How to Bring the Sciences into Democracy.* Trans. Catherine Porter. Cambridge, Mass.: Harvard University Press.
Massumi, Brian. 2002. "The Political Economy of Belonging and the Logic Of Relation." In *Parables for the Virtual,* 68–88. Durham, N.C.: Duke University Press.
Starhawk. 1997. *Dreaming the Dark.* Boston: Beacon Press.
Stengers, Isabelle. 1996–97. *Cosmopolitiques.* 7 vols. Paris: La Découverte.
———. 2000. *The Invention of Modern Science.* Trans. Daniel W. Smith. Minneapolis: University of Minnesota Press.
———. 2006. *La Vierge et le Neutrino: Les scientifiques dans la tourmente.* Paris: Empêcheurs de Penser en Rond.
Weinberg, Stephen. 1993. *Dreams of a Final Theory.* London: Vintage.
Whitehead, Alfred N. 1968. *Modes of Thought.* New York: Free Press.

2 Thing-Power

JANE BENNETT

> I must let my senses wander as my thought, my eyes see without looking. . . . Go not to the object; let it come to you.
>
> —Henry Thoreau, *The Journal of Henry David Thoreau*

> It is never we who affirm or deny something of a thing; it is the thing itself that affirms or denies something of itself in us.
>
> —Baruch Spinoza, *Short Treatise II*

IN THE WAKE OF FOUCAULT'S DEATH IN 1984, there was an explosion of scholarship on the body and its social construction, on the operations of biopower. These genealogical (in the Nietzschean sense) studies exposed the various micropolitical and macropolitical techniques through which the human body was disciplined, normalized, sped up and slowed down, gendered, sexed, nationalized, globalized, rendered disposable, or otherwise composed. The initial insight was to reveal how cultural practices produce what is experienced as natural, but many theorists also insisted on the *material recalcitrance* of such cultural productions.[1] Though gender, for example, was a congealed bodily effect of historical norms and repetitions, its status as artifact does *not* imply an easy susceptibility to human understanding, reform, or control. The point was that cultural forms are themselves powerful, material assemblages with *resistant force.*

In what follows, I too will feature the negative power or recalcitrance of things. But I will also seek to highlight a positive, productive power of their own. And instead of collectives conceived primarily as conglomerates of *human* designs and practices ("discourse"), I will highlight the active role of *nonhuman* materials in public life. In short, I will try to give voice to a *thing-power.* As W. J. T. Mitchell (2005, 156–57) notes, "objects are the way things appear to a subject—that is, with a name, an identity, a gestalt or stereotypical template . . . Things, on the other hand, . . . [signal] the moment when the object becomes the Other, when the sardine can looks back, when the mute idol speaks, when the subject experiences the object as uncanny and feels the need for what Foucault calls 'a metaphysics of the object, or, more exactly, a metaphysics of that never objectifiable depth from which objects rise up toward our superficial knowledge.'"

Thing-Power, or the Out-Side

Spinoza ascribes to bodies a peculiar vitality: "each thing *[res]*, as far as it can by its own power, strives *[conatur]* to persevere in its own being" (Spinoza 1982, pt. 3, proposition 6). *Conatus* names an "active impulsion" (Mathews 2003, 8) or trending tendency to persist. Although Spinoza distinguishes the human body from other bodies by noting that its distinctive "virtue" consists in "nothing other than to live by the guidance of reason" (Spinoza 1982, pt. 4, proposition 37, scholium 1), every nonhuman body shares with every human body a conative nature (and thus a "virtue" appropriate to its material configuration). Conatus names a power present in *every* body: "Any thing whatsoever, whether it be more perfect or less perfect, will always be able to persist in existing with that same force whereby it begins to exist, so that in this respect all things are equal" (Spinoza 1982, 154). Even a falling stone, writes Spinoza (1995, epistle 58), "is endeavoring, as far as in it lies, to continue in its motion."[2] As Nancy Levene (2004, 3) notes, "Spinoza continually stresses this continuity between human and other beings," for "not only do human beings not form a separate imperium unto themselves; they do not even command the imperium, nature, of which they are a part."[3]

The idea of thing-power bears a family resemblance to Spinoza's conatus as well as to what Thoreau called the Wild or to that uncanny presence that met him in the Concord woods and atop Mt. Ktaadn and also resided in/as that monster called the railroad and that alien called his Genius. Wildness was a not-quite-human force that addled and altered human and other bodies. It named an irreducibly strange dimension of matter, an *out-side*. Thing-power is also kin to what Hent de Vries (2006, 42), in the context of political theology, called "the absolute" or that "intangible and imponderable" recalcitrance. Though the absolute is often equated with God, de Vries defines it more open-endedly as "that which tends to loosen its ties to existing contexts" (6). This definition makes sense when we look at the etymology of *absolute: ab* (off) + *solver* (to loosen). The absolute is that which is *loosened off*, on the loose, on the outside. When, for example, a Catholic priest performs the act of ab-solution, he is the vehicle of a divine agency that *loosens* sins from their attachment to a particular soul: sins now stand apart, displaced foreigners living a strange, "impersonal life" of their own. When de Vries speaks of the absolute, then, he tries to point to what no speaker could possibly see, that is, a some-thing that is not an object of knowledge is detached or radically free from representation and thus is no-thing at all—nothing but the force or effectivity of this detachment.

De Vries's notion of the absolute, like the thing-power I will seek to express, seeks to acknowledge that which refuses to dis-solve completely into the milieu of human knowledge. But there is also a difference in emphasis between De Vries and me. De Vries conceives this exteriority, this out-side, primarily as an epistemological limit: in the presence of the absolute, we cannot *know*. It is from human thinking that the absolute has detached; the absolute names the limits of *intelligibility*. De Vries's formulations thus, naturally enough, give priority to humans as knowing bodies, while overlooking things and what *they* can do. The notion of thing-power aims instead to attend to the it as actant; I will try, impossibly, to name the moment of independence (from subjectivity) possessed by things, a moment that must be there since things do in fact affect other bodies, enhancing or weakening their power. I will shift from the language of epistemology to ontology, from a focus on an elusive recalcitrance hovering between immanence and transcendence (the absolute) to an active, earthy, not-quite-human capaciousness (vital materiality). I will try to give voice to a vitality intrinsic to materiality, in the process absolving matter from its long history of attachment to automatism or mechanism.[4]

The strangely vital things that will rise up to meet us in this chapter—a dead rat, a plastic cap, a spool of thread—are characters in a speculative onto-story. The tale hazards an account of materiality, even though it is both too alien and too close to see clearly and even though linguistic means are inadequate to the task. The story will highlight the extent to which human being and thinghood overlap, the extent to which the "us" and the "it" slip-slide into each other. One moral of the story is that we are also nonhuman and that things, too, are vital players in the world. The hope is that the story will enhance receptivity to the "impersonal life" that surrounds and infuses us, generate a more subtle awareness of the complicated web of dissonant connections between bodies, and enable more strategic interventions into that ecology.

Thing-Power I: Debris

On a sunny Tuesday morning on June 4, in the grate over the storm drain to the Chesapeake Bay in front of Sam's Bagels on Cold Spring Lane in Baltimore, there was

- one large men's black plastic work glove
- one dense mat of oak pollen
- one unblemished dead rat
- one white plastic bottle cap
- one smooth stick of wood

Glove, pollen, rat, cap, stick. As I encountered these items, they shimmied back and forth between debris and thing—between, on one hand, stuff to ignore, except insofar as it betokened human activity (the workman's efforts, the litterer's toss, the rat-poisoner's success) and, on the other hand, stuff that commanded attention in its own right, as an existent in excess of its association with human meanings, habits, or projects. In the second moment, stuff exhibited its thing-power: it issued a call, even if I did not quite understand what it was saying. At the very least, it provoked affects in me: I was repelled by the dead (or was it merely sleeping?) rat and dismayed by the litter, but I also felt something else: a nameless awareness of the impossible singularity of *that* rat, that configuration of pollen, that otherwise utterly banal, mass-produced plastic water-bottle cap.

I was struck by what Stephen Jay Gould (2002, 1338) called the "excruciating complexity and intractability" of nonhuman bodies. But in being *struck*, I realized that the capacity of these bodies was not restricted to a passive "intractability" and also included the ability to make things happen, to produce effects. When the materiality of the glove, the rat, the pollen, the bottle cap, the stick started to shimmer and spark, it was in part because of the contingent tableau that they formed with each other, with the street, with the weather that morning, with me. For had the sun not glinted on the black glove, I might not have seen the rat; had the rat not been there, I might not have noted the bottle cap; and so on. But they *were* all there just as they were, and so I caught a glimpse of an energetic vitality inside each of these things, things that I generally conceived as inert. In this assemblage, *objects* appeared as *things*, that is, as vivid entities not entirely reducible to the contexts in which (human) subjects set them, never entirely exhausted by their semiotics. In my encounter with the gutter on Cold Spring Lane, I glimpsed a culture of things irreducible to the culture of objects.[5] I achieved, for a moment, what Thoreau had made his life's goal: to be able "to be surprised by what we see" (Dumm 1999, 7).[6]

This window onto an eccentric out-side was made possible by the fortuity of that particular assemblage but also by a certain anticipatory readiness on my in-side, by a perceptual style open to the appearance of thing-power. For I came on the glove-pollen-rat-cap-stick with Thoreau in my head, who had encouraged me to practice "the discipline of looking always at what is to be seen" (Thoreau 1973, 111),[7] with Spinoza's (1982, proposition 13, scholium 72) claim that all things are "animate, albeit in different degrees" and also with Merleau-Ponty (1981), whose *Phenomenology of Perception* had disclosed for me "an immanent or incipient significance in the living body [which] extends, . . . to the whole sensible world" and which had shown me how "our gaze, prompted by the experience of our own body, will discover in all other 'objects' the miracle of expression" (197).

As I have already noted, the items on the ground that day were vibratory, at one moment disclosing themselves as dead stuff and at the next as a live presence. Junk then claimant, inert matter then live wire. It hit me then in a visceral way how American materialism, which requires buying ever-increasing numbers of products purchased in ever-shorter cycles, is antimateriality.[8] The sheer volume of commodities, and the hyperconsumptive necessity of junking them to make room for new ones, conceals the vitality of matter. In *The Meadowlands,* a late-twentieth-century Thoreauian travelogue of the New Jersey garbage hills outside Manhattan, Robert Sullivan describes the vitality that persists even in trash:

> The . . . garbage hills are alive. . . . There are billions of microscopic organisms thriving underground in dark, oxygen-free communities. . . . After having ingested the tiniest portion of leftover New Jersey or New York, these cells then exhale huge underground plumes of carbon dioxide and of warm moist methane, giant stillborn tropical winds that seep through the ground to feed the Meadowlands' fires, or creep up into the atmosphere, where they eat away at the . . . ozone. . . . One afternoon I . . . walked along the edge of a garbage hill, a forty-foot drumlin of compacted trash that owed its topography to the waste of the city of Newark. . . . There had been rain the night before, so it wasn't long before I found a little leachate seep, a black ooze trickling down the slope of the hill, an espresso of refuse. In a few hours, this stream would find its way down into the . . . groundwater of the Meadowlands; it would mingle with toxic streams. . . . But in this moment, here at its birth, . . . this little seep was pure pollution, a pristine stew of oil and grease, of cyanide and arsenic, of cadmium, chromium, copper, lead, nickel, silver, mercury, and zinc. I touched this fluid—my fingertip was a bluish caramel color—and it was warm and fresh. A few yards away, where the stream collected into a benzene-scented pool, a mallard swam alone. (Sullivan 1998, 96–97)

Sullivan reminds that a vital materiality can never really be thrown "away." For it continues its activities even as a discarded or unwanted commodity. For Sullivan that day, as for me on that June morning, thing-power rose from a pile of trash. Flower Power, Black Power, Girl Power. *Thing Power:* the curious ability of inanimate things to animate, to act, to produce effects dramatic and subtle.

Thing-Power II: Odradek's Nonorganic Life

A dead rat, some oak pollen, and a stick of wood stopped me in my tracks. But so did the plastic glove and bottle cap: thing-power arises from bodies inorganic as well as organic. In support of this contention, Manuel De Landa notes how even inorganic matter can "self-organize":

> inorganic matter-energy has a wider range of alternatives for the generation of structure than just simple phase transitions. . . . In other words, even the humblest forms of matter and energy have the potential for *self-organization* beyond the relatively simple type involved in the creation of crystals. There are, for instance, those coherent waves called solitons which form in many different types of materials, ranging from ocean waters (where they are called tsunamis) to lasers. Then there are . . . stable states (or attractors), which can sustain coherent cyclic activity. . . . Finally, and unlike the previous examples of nonlinear self-organization where true innovation cannot occur, there [are] . . . the different combinations into which entities derived from the previous processes (crystals, coherent pulses, cyclic patterns) may enter. When put together, these forms of spontaneous structural generation suggest that inorganic matter is much more variable and creative than we ever imagined. And this insight into matter's inherent creativity needs to be fully incorporated into our new materialist philosophies. (De Landa 2000, 16)

Here I would like to draw attention to a literary dramatization of this idea: to Odradek, the protagonist of Kafka's short story "Cares of a Family Man." Odradek is a spool of thread who/that can run and laugh; this animate wood exercises an impersonal form of vitality. De Landa speaks of a "spontaneous structural generation," which happens, for example, when chemical systems at far-from-equilibrium states inexplicably choose one path of development rather than another. Like these systems, the material configuration that is Odradek straddles the line between inert matter and vital life.

That is why Kafka's narrator has trouble assigning Odradek to an ontological category. Is Odradek a cultural artifact, a tool of some sort? Perhaps, but if so, its purpose is obscure:

> It looks like a flat star-shaped spool of thread, and indeed it does seem to have thread wound upon it; to be sure, these are only old, broken-off bits of thread, knotted and tangled together, of the most varied sorts and colors. . . . One is tempted to believe that the creature once had some sort of intelligible shape and is now only a broken-down remnant. Yet this does not seem to be the case; . . . nowhere is there an unfinished or unbroken surface to suggest anything of the kind: the whole thing looks senseless enough, but in its own way perfectly finished. (Kafka 1983, 428)

Or perhaps Odradek is more a subject than an object—an organic creature, a little person? But if so, his/her/its embodiment seems rather unnatural: from the center of Odradek's star there protrudes a small wooden crossbar, and "by means of this latter rod . . . and one of the points of the star . . . the whole thing can stand upright as if on two legs."

On one hand, like an active organism, Odradek appears to move deliberately (he is "extraordinarily nimble") and to speak intelligibly:

> He lurks by turns in the garret, the stairway, the lobbies, the entrance hall. Often for months on end he is not to be seen; then he has presumably moved into other houses; but he always comes faithfully back to our house again. Many a time when you go out of the door and he happens just to be leaning directly beneath you against the banisters you feel inclined to speak to him. Of course, you put no difficult questions to him, you treat him—he is so diminutive that you cannot help it—rather like a child. "Well, what's your name?" you ask him. "Odradek," he says. "And where do you live?" "No fixed abode," he says and laughs. (Kafka 1983, 428)

And yet, on the other hand, like an inanimate object, Odradek's so-called laughter "has no lungs behind it" and "sounds rather like the rustling of fallen leaves. And that is usually the end of the conversation. Even these answers are not always forthcoming; often he stays mute for a long time, as wooden as his appearance" (Kafka 1983, 428).

Wooden yet lively, verbal yet vegetal, alive and inert, Odradek is ontologically multiple. He/it is a vital materiality and exhibits what Deleuze described as the persistent "hint of the animate in plants, and of the vegetable in animals" (Deleuze 1991, 95). The Russian scientist Vladimir Ivanovich Vernadsky (1863–1945), who also refused any sharp distinction between life and matter, defined organisms as "special, distributed forms of the common mineral, water. . . . Emphasizing the continuity of watery life and rocks, such as that evident in coal or fossil limestone reefs, Vernadsky noted how these apparently inert strata are 'traces of bygone biospheres'" (Margulis and Sagan 1995, 50). Odradek exposes this continuity of watery life and rocks; he brings to the fore the becoming of things.

Thing-Power III: Legal Actants

I may have met a relative of Odradek while serving on a jury, again in Baltimore, for a man on trial for attempted homicide. It was a small glass vial with an adhesive-covered metal lid: the gunpowder residue sampler. This object/witness had been dabbed on the accused's hand hours after the shooting and now offered to the jury its microscopic evidence that the hand had either fired a gun or been within three feet of a gun firing. The sampler was shown to the jury several times by expert witnesses, and with each appearance it exercised more force, until it became vital to the verdict. This composite of glass, skin cells, glue, words, laws, metals, and human emotions had became an *actant.* An actant is Latour's term for a source of action; an actant can be human or not, or, most likely, a combination of both. Latour defines it as "something that acts or to which activity is granted by others. It implies no special motivation of human individual actors, nor of humans in general" (Latour 1997). An actant is neither an object nor a subject, but an "intervener," akin to the Deleuzean idea of a "quasi-causal operator" (De Landa 2003, 123). An operator is that which, by virtue of its particular location in an assemblage and the fortuity of being in the right place *at the right time,* makes the difference, makes things happen, becomes the decisive force which catalyzes an event.

Actants and operators are substitute words for what, in a more subject-centered vocabulary, are called agents. Agentic capacity is now seen as differentially distributed across a wider range of ontological types. This idea is also expressed in the notion of "deodand," a figure of English law from about 1200 until it was abolished in 1846. In cases of accidental death or injury to a human, the nonhuman actant, for example, the carving knife that fell into human flesh or the carriage that trampled the leg of a pedestrian—became deodand (literally, "that which must be given to God"). In recognition of *its* peculiar efficacy (a power that is less masterful than agency but more active than recalcitrance), the deodand, a materiality "suspended between human and thing" (Tiffany 2001, 74),[9] was surrendered to the Crown to be used (or sold) to compensate for the harm done. According to William Pietz, "any culture must establish some procedure of compensation, expiation, or punishment to settle the debt created by unintended human deaths whose direct cause is not a morally accountable person, but a nonhuman material object. This was the issue thematized in public discourse by . . . the law of deodand" (Pietz 1997, 97).

There is, of course, a difference between the knife that impales and the man impaled, between the technician who dabs the sampler and the sampler, between the array of items in the gutter of Cold Spring Lane and me, the narrator of their vitality. But I agree with John Frow that this difference "needs to be flattened, read horizontally as a juxtaposition rather than vertically as a hierarchy of being. It's a feature of our world that we can and do distinguish . . . things from persons. But the sort of world we live in makes it constantly possible for these two sets of kinds to exchange properties" (Frow 2001, 283). And to note this fact explicitly, which is also to begin to experience the relationship between persons and other materialities more horizontally, is to take a step toward a more ecological sensibility.

Thing-Power IV: Walking, Talking Minerals

Odradek, a gunpowder residue sampler, and some junk on the street can be fascinating to people and can seem to come alive. But is this evanescence a property of the stuff or of people? Was the "thing-power" of the debris I encountered but a function of the subjective and intersubjective connotations, memories, and emotions that had accumulated around my ideas of these items? Was the real agent of my temporary immobilization on the street that day *humanity,* that is, the cultural meanings of "rat," "plastic," "wood" in conjunction with my own idiosyncratic biography? It could be. But what if the swarming activity inside my head was *itself* an instance of the vital materiality that also constituted the trash?

I have been trying to raise the volume on the vitality of materiality per se, pursuing this task so far by focusing on nonhuman bodies, by, that is, depicting them as actants rather than objects. But the case for matter as active needs also to readjust the status of human actants: not by denying humanity's awesome, awful powers but by presenting these powers as evidence of our own constitution as vital materiality. In other words, human power is itself a kind of thing-power. At one level, this claim is uncontroversial: it is easy to acknowledge that humans are composed of various material parts (the minerality of our bones, or the metal of our blood, or the electricity of our neurons). But it is more challenging to conceive of these materials as lively and self-organizing than as the passive or mechanical means under the direction of something nonmaterial, that is, an active soul or mind.

But perhaps the claim to a vitality intrinsic to matter itself becomes more plausible if one takes a long view of time. If one adopts the perspective of evolutionary rather than biographical time, for example, a mineral efficacy becomes visible. Here is De Landa's account of the emergence of our bones:

> Soft tissue (gels and aerosols, muscle and nerve) reigned supreme until 5000 million years ago. At that point, some of the conglomerations of fleshy matter-energy that made up life underwent a sudden *mineralization,* and a new material for constructing living creatures emerged: bone. It is almost as if the mineral world that had served as a substratum for the emergence of biological creatures was reasserting itself. (De Landa 2000, 26)

Mineralization names the creative agency by which bone was produced, and bones then "made new forms of movement control possible among animals, freeing them from many constraints and literally setting them into motion to conquer every available niche in the air, in water, and on land" (26–27). In the long and slow time of evolution, then, mineral material appears as the mover and shaker, the active power, and the human being, with his much-lauded capacity for self-directed action, appears as *its* product.[10] Vernadsky seconds this view in his description of humankind as a particularly potent mix of minerals:

> What struck [Vernadsky] most was that the material of Earth's crust has been packaged into myriad moving beings whose reproduction and growth build and break down matter on a global scale. People, for example, redistribute and concentrate oxygen . . . and other elements of Earth's crust into two-legged, upright forms that have an amazing propensity to wander across, dig into and in countless other ways alter Earth's surface. *We are walking, talking minerals.* (Margulis and Sagan 1995, 49; italics added)

Kafka, De Landa, and Vernadsky suggest that human individuals are themselves composed of vital materials, that our powers are thing-power. These vital materialists do not claim that there are no differences between humans and bones, only that there is no need to describe these differences in a way that places humans at the ontological center or hierarchical apex. Humanity can be distinguished, instead, as Jean-Francois Lyotard suggests, as a *particularly rich and complex* collection of materials: "humankind is taken for a complex material system; consciousness, for an effect of language; and language for a highly complex material system" (Lyotard 1997, 98). Richard Rorty similarly defines humans as very complex animals rather than as animals "with an extra added ingredient called 'intellect' or 'the rational soul'" (Rorty 1995, 199).

The fear is that in failing to affirm human uniqueness, such views authorize the treating of people as mere things, in other words, that a strong distinction between subjects and objects is needed to prevent the instrumentalization of humans. Yes, such critics continue, objects possess a certain power of action (as when bacteria or pharmaceuticals enact hostile or symbiotic projects inside the human body), and yes, some subject-on-subject objectifications are permissible (as when persons consent to use and be used as a means to sexual pleasure), but the *ontological* divide between persons and things must remain lest one have no *moral* grounds for privileging human over germ or for condemning pernicious forms of human-on-human instrumentalization (as when powerful humans exploit illegal, poor, young, or otherwise weaker humans).

How can the vital materialist respond to this important concern? First, he can respond by acknowledging that at times the framework of subject versus object has worked to prevent or ameliorate human suffering and to promote human happiness or well-being, and second, by noting that its successes come at the price of an instrumentalization of nonhuman nature that itself can be unethical and undermine long-term human interests. Third, he must point out that the Kantian imperative to treat humanity always as an end in itself and never merely as a means does not have a stellar record of success in preventing human suffering or promoting human well-being: it is important to raise the question of its actual, historical efficacy to open up space for forms of ethical practice that do not rely on the image of an intrinsically *hierarchical* order of things. Here the materialist speaks of promoting healthy and enabling instrumentalizations rather than of treating people as ends in themselves because to face up to the compound nature of the human self is to find it difficult even to make sense of the notion of a single end in itself. What instead appears is a swarm of competing ends being pursued simultaneously in each individual, some of which are healthy to the whole, some of which are not. Here the vital materialist, taking a cue from Nietzsche's and Spinoza's ethics, favors physiological (healthy and enabling) over moral terms (dignity, purposiveness, divinity) because she fears that moralism can itself become a source of unnecessary human suffering.[11]

We are now in a position to name that other way to promote human health and happiness: *to raise the status of the materiality of which we are composed.* Each human is a heterogeneous compound of wonderfully, dangerously vibrant matter. If matter itself is lively, then not only is the difference between subjects and objects minimized, but the status of the shared materiality of all things is elevated. All bodies become more than mere objects, as the thing-powers of resistance and protean agency are brought into sharper relief. Vital materialism would thus set up a kind of safety net for those humans who are now, in a world where Kantian morality is the standard, routinely made to suffer because they do not conform to a particular (Euro-American, bourgeois, theocentric, or other) model of personhood. The ethical aim becomes to distribute value more generously, to bodies as such. Such a newfound attentiveness to matter and its powers will not solve the problem of human exploitation or oppression, but it can inspire a greater sense of the extent to which all bodies are kin in the sense of being inextricably enmeshed in a dense network of relations. And in a knotted world of vibrant matter, to harm one section of the web may very well be to harm oneself. Such an enlightened or expanded notion of self-interest *is good for humans.* A vital materialism does not reject self-interest as a motivation for ethical behavior, though it does seek to cultivate a broader definition of self and of interest.

Thing-Power V: Thing-Power and Adorno's Nonidentity

But perhaps the very idea of thing-power or vibrant matter claims too much: to know more than it is possible to know. Or, to put the criticism in Theodor Adorno's terms, does it exemplify the violent hubris of Western philosophy, a tradition that has consistently failed to mind the gap between concept and reality, object and thing? For Adorno, this gap is ineradicable, and the very most that can be said with confidence about the thing is that it eludes capture by the concept that there is always a "nonidentity" between it and any representation. And yet, as I shall argue, even Adorno continues to seek a way to access—however darkly, crudely, or fleetingly—this outside. One can detect a trace of this longing in the following quotation from *Negative Dialectics:* "What we may call the thing itself is not positively and immediately at hand. He who wants to know it must think more, not less" (Adorno 1973, 189). Adorno clearly rejects the possibility of any direct, sensuous apprehension ("the thing itself is not positively and immediately at hand"), but he does not reject all modes of encounter, for there is one mode, "thinking more, not less," that holds promise.

In this section, I will explore some of the affinities between Adorno's nonidentity and my thing-power and, more generally, between his "specific materialism" and a vital materialism.

Nonidentity is the name Adorno gives to that which is not subject to knowledge but is instead "heterogeneous" to all concepts. This elusive force is not, however, wholly outside of human experience, for Adorno describes nonidentity as a presence that acts on us: we knowers are haunted, he says, by a painful, nagging feeling that something is being forgotten or left out. This discomfiting sense of the inadequacy of representation remains no matter how refined or analytically precise one's concepts become. *Negative dialectics* is the method Adorno designs to teach us how to *accentuate* this discomforting experience and how to give it a meaning. When practiced correctly, negative dialectics will render the static buzz of nonidentity into a powerful reminder that "objects do not go into their concepts without leaving a remainder" and thus that life will always exceed our knowledge and control. The ethical project par excellence, as Adorno sees it, is to keep remembering this and to learn how to accept it. Only then can we stop raging against a world that refuses to offer us the "reconcilement" that we, according to Adorno, crave (Adorno 1973, 5).[12]

For the vital materialist, however, the starting point of ethics is less the acceptance of the impossibility of "reconcilement" and more the recognition of human participation in a shared, vital materiality. We *are* vital materiality, and we are surrounded by it, though we do not always see it that way. The ethical task at hand here is to cultivate the ability to discern nonhuman vitality, to become perceptually open to it. In a parallel manner, Adorno's "specific materialism" also recommends a set of practical techniques for training oneself to better detect and accept nonidentity. Negative dialectics is, in other words, the pedagogy inside Adorno's materialism.

This pedagogy includes intellectual as well as aesthetic exercises. The intellectual practice consists in the attempt to make the very process of conceptualization an explicit object of thought. The goal here is to become more cognizant of the fact that conceptualization automatically obscures the inadequacy of its concepts. Adorno believes that critical reflection can expose this cloaking mechanism and that the exposure will intensify the felt presence of nonidentity. The treatment is homeopathic: we must develop a *concept* of nonidentity to cure the hubris of conceptualization. The treatment can work because, however distorting, concepts still "refer to nonconceptualities." This is "because concepts on their part are moments of the reality that requires their formation" (Adorno 1973, 12). Concepts can never provide a clear view of things in themselves, but the "discriminating man," who "in the matter and its concept can distinguish even the infinitesimal, that which escapes the concept," can do a better job of gesturing toward them (45). Note that the discriminating man (adept at negative dialectics) both subjects his conceptualizations to second-order reflection and also pays close *aesthetic* attention to the object's "qualitative moments," for these open a window onto nonidentity (43).

A second technique of the pedagogy is to exercise one's utopian imagination. The negative dialectician should imaginatively re-create what has been obscured by the distortion of conceptualization: "the means employed in negative dialectics for the penetration of its hardened objects is possibility—the possibility of which their reality has cheated the objects and which is nonetheless visible in each one" (Adorno 1973, 52). Nonidentity resides in those denied possibilities, in the invisible field that surrounds and infuses the world of objects.

A third technique is to admit a "playful element" into one's thinking and to be willing to play the fool. The negative dialectician "knows how far he remains from" knowing nonidentity, "and yet he must always talk as if he had it entirely. This brings him to the point of clowning. He must not deny his clownish traits, least of all since they alone can give him hope for what is denied him" (Adorno 1973, 14).

The self-criticism of conceptualization, a sensory attentiveness to the qualitative singularities of the object, the exercise of an unrealistic imagination, and having the courage of a clown: by means of such practices, one might replace the rage against nonidentity with a respect for it, a respect that chastens our will to mastery. That rage is for Adorno the driving force behind interhuman acts of cruelty and violence. Adorno goes even further to suggest that negative dialectics can transmute the anguish of nonidentity into a will to ameliorative political action: the thing thwarts our desire for conceptual and practical mastery, and this refusal angers us; but it also offers us an ethical injunction, according to which "suffering ought not to be, . . . things should be different. 'Woe speaks: "Go."' Hence the convergence of specific materialism with criticism, with social change in practice" (Adorno 1973, 202–3).[13]

Adorno founds his ethics on an intellectual and aesthetic attentiveness that, though it will always fail to see its object clearly, nevertheless has salutory effects on the bodies straining to see. Adorno willingly plays the fool by questing after what I would call thing-power but that he calls "the preponderance of the object" (Adorno 1973, 183). Humans encounter a world where nonhuman materialities have power, a power that the "bourgeois I," with its pretensions to autonomy, denies.[14] It is at this point that Adorno identifies negative dialectics as a materialism: it is only "by passing to the object's preponderance that dialectics is rendered materialistic" (192).

Adorno dares to affirm something like thing-power, but he does not want to play the fool for *too* long. He is quick—too quick from the point of view of the vital materialist—to remind the reader that objects are always "entwined" with human subjectivity and that he has no desire "to place the object on the orphaned royal throne once occupied by the subject. On that throne the object would be nothing but an idol" (Adorno 1973, 181). Adorno is reluctant to say too much about nonhuman vitality, for the more said, the more it recedes from view. Nevertheless, Adorno does try to *attend* somehow to this reclusive reality by means of a negative dialectics. Negative dialectics has an affinity with negative *theology:* negative dialectics honors nonidentity as one would honor an unknowable god; Adorno's "specific materialism" includes the possibility that there is divinity behind or within the reality that withdraws. Adorno rejects any naive picture of transcendence, such as that of a loving God who designed the world ("Metaphysics cannot rise again" after Auschwitz), but the desire for transcendence cannot, he believes, be eliminated: "nothing could be experienced as truly alive if something that transcends life were not promised also. . . . The transcendent is, and it is not" (404, 375).[15] Adorno honors nonidentity as an *absent* absolute, as a messianic promise.[16]

Adorno struggles to describe a force that is *material* in its resistance to human concepts but *spiritual* insofar as it might be a dark promise of an absolute-to-come. A vital materialism is more thoroughly nontheistic in presentation: the out-side has no messianic promise.[17] But a philosophy of nonidentity and a vital materialism nevertheless share an urge to cultivate a more careful attentiveness to the out-side.

The Naive Ambition of Vital Materialism

Adorno reminds us that humans can experience the out-side only indirectly, only through vague, aporetic, or unstable images and impressions. But when he says that even distorting concepts still "refer to nonconceptualities, because concepts on their part are moments of the reality that requires their formation" (Adorno 1973, 12), Adorno also acknowledges that human experience nevertheless includes encounters with an out-side that is active, forceful, and (quasi)independent. This out-side can operate at a distance from our bodies, or it can operate as a foreign power internal to them, as when we feel the discomfort of nonidentity or hear the nay-saying voice of Socrates' demon or are moved by what Lucretius called that "something in our chest capable of fighting and resisting," a *thumos* "brought about by a tiny swerve of atoms" (Lucretius 1995, 128). There is a strong tendency among modern, secular, well-educated humans to refer such signs back to a human agency conceived as its ultimate source. This impulse toward cultural, linguistic, or historical constructivism, which interprets any expression of thing-power as an effect of culture and the play of human powers, politicizes moralistic and oppressive appeals to "nature." And that is a good thing. But the constructivist response to the world would also tend to obscure from view whatever thing-power there may be. There is, then, something to be said for moments of methodological naivete, for the postponement of a genealogical critique of objects.[18] This delay might render manifest a subsistent world of nonhuman vitality. To "render manifest" is both to receive and to participate in the shape given to that which is received. What is manifest arrives through humans but not entirely because of them.

A vital materialist will try to linger, then, in those moments when she finds herself fascinated by objects, taking them as clues to the material vitality that she shares with them. This sense of a strange and incomplete commonality with the outside may induce her to treat nonhumans—animals, plants, earth, even artifacts and commodities—more carefully, more strategically, more ecologically. But how to develop this capacity for naiveté? One tactic might be to revisit and become temporarily infected by discredited philosophies of nature, risking "the taint of superstition, animism, vitalism, anthropomorphism, and other premodern attitudes" (Mitchell 2005, 149). To this end, let me make a brief foray into the ancient atomism of Lucretius, Roman devotee of Epicurus.

Lucretius tells of bodies falling in a void, bodies that are not lifeless stuff but matter on the go, entering and leaving assemblages, swerving into each other:

> *at times quite undetermined and at undetermined spots they push a little from their path:* yet only just so much as you could call a change of trend. [For if they did not] . . . swerve, all things would fall downwards through the deep void like drops of rain, nor could collision come to be, nor a blow brought to pass for the primordia: so nature would never have brought anything into existence. (Lucretius 1995, 216)[19]

Althusser described this as a "materialism of the encounter," according to which political events are born from chance meetings of atoms (Althusser 2006, 169).[20] A primordial swerve says that the world is not determined, that an element of chanciness resides at the heart of things, but it also affirms that so-called inanimate things have a life, that deep within is an inexplicable vitality or energy, a moment of independence from and resistance to us and other bodies—a kind of thing-power.

The rhetoric of *De Rerum Natura* is realist, speaking in an authoritative voice, claiming to describe a nature that preexists and outlives us: here are the smallest constituent parts of being ("primordial"), and here are the principles of association governing them.[21] It is easy to criticize this realism: Lucretius quests for the thing itself, but there is no there there—or, at least, no way for us to grasp it or know it, for the thing is always already humanized, its object-status arises the very instant something comes into our awareness. Adorno levels this charge explicitly against Heidegger's phenomenology, which Adorno interprets as a "realism" that "seeks to breach the walls which thought has built around itself, to pierce the interjected layer of subjective positions that have become a second nature." Heidegger's aim, that is, "to philosophize formlessly, so to speak, purely on the ground of things" (Adorno 1973, 78),[22] is for Adorno as futile, and it is productive of a violent "rage" against nonidentity.[23]

But Lucretius's poem, like Kafka's stories, Sullivan's travelogue, Vernadsky's speculations, and my account of the gutter of Cold Spring Lane, does offer this potential benefit: it can direct sensory, linguistic, and imaginative attention toward a material vitality. The advantage of such tales, with their ambitious naiveté, is that though they "disavow . . . the tropological work, the psychological work, and the phenomenological work entailed in the human production of materiality," they do so "in the name of *avowing* the force of questions that have been too readily foreclosed by more familiar fetishizations: the fetishization of the subject, the image, the word."[24]

There is a rich archive in Euro-American political theory for thinking about materiality, from Plato's depiction of the ghostly objects in the Cave to Marx's invention of a distinctly *historical* materialism to twentieth-century feminist theories of the body. In many of these sources, there is, quite naturally, a focus on the power of humans—their minds, their bodies, and the interactions between the two powers. To the extent that politics is figured as a *human* field of action, the technophysical "stuff" of politics has tended to register in political theory as a set of (enabling or constraining) material *conditions* for what is imagined as the ultimate stuff of politics: humans acting in concert. I have in this essay drawn on some (quasi)canonical writers, in particular, Spinoza, Kafka, and Adorno, to emphasize, even overemphasize, the agentic contributions of nonhuman forces (operating in "nature," in the human body, and in human artifacts). Such *over*-emphasis, it seems to me, is needed to counter the powerfully narcissistic reflex of human language and thought. To think more carefully about the stuff of politics, then, a bit of anthropomorphism—the idea that human agency has some echoes in nonhuman nature—may be useful. Politics *does* involve humans acting in concert, but in concert with quite an ontologically diverse range of actors.

Notes

A version of this essay also appears as chapter 1 of Jane Bennett, *Vibrant Matter* (Durham, N.C.: Duke University Press, 2010), 1–19. The version printed here originally appeared as Jane Bennett, "The Force of Things," *Political Theory* 32, no. 3 (2004): 347–72. Copyright 2004, Sage Publications.

1 There is too much good work here in feminist theory, queer studies, and cultural studies to cite. The three volumes of Michel Feher (1989) offer one map of the terrain. See also Butler (1993, 1998), Brown (1995), Ferguson (1991), and Gatens (1996).

2 Spinoza links, in this famous letter, his theory of *conatus* to a critique of the notion of human free will: "Now this stone, since it is conscious only of its endeavor *[conatus]* and is not at all indifferent, will surely think that it is completely free, and that it continues in motion for no other reason than it so wishes. This, then, is that human freedom which all men boast of possessing, and which consists solely in this, that men are conscious of their desire and unaware of the causes by which they are determined" (Spinoza 1995, epistle 58). Hasana Sharp (2007, 740) argues that the analogy between humans and stones "is not as hyperbolic as it seems at first glance. For Spinoza, all beings, including stones . . . include a power of thinking that corresponds exactly to the power of their bodies to be disposed in different ways, to act and be acted upon. . . . Likewise every being, to the extent that it preserves its integrity amidst infinitely many other beings, as a stone surely does, is endowed with . . . a desire

to . . . preserve and enhance its life to the extent that its nature allows."

3 Yitshak Melamed goes further to say that "since the doctrine of the conatus . . . provide[s] the foundations for Spinoza's moral theory, it seems likely that we could even construct a moral theory for hippopotamuses and rocks" (Melamed 2010, n59).

4 De Vries (2006) seems to affirm this association when he wonders whether Spinoza's picture of interacting, conatus-driven *bodies* could possibly account for the creative emergence of the new: "it would seem that excess, gift, the event . . . have no place here" (22). Why? Because the only plausible locus of creativity is, for de Vries, one that is "quasi-spiritual," hence Spinoza's second attribute of God/Nature, i.e., *thought* or ideas. But what if materiality itself harbors creative vitality?

5 On the effectivity of trash, see Tim Edensor's (2005) fascinating work.

6 See Dumm (1999, 7) for a subtle reckoning with the "obscure power of the ordinary." My attempt to speak on behalf of "things" is a companion project to Dumm's attempt to mine the ordinary as a potential site of resistance to conventional and normalizing practices.

7 Thoreau (1949, 313) trained his gaze on things with faith that "the perception of surfaces will always have the effect of miracle to a sane sense."

8 For a good analysis of the implications of the trash-and-waste culture for democracy, see Buell and DeLuca (1996).

9 Tiffany (2001) draws an analogy between riddles and materiality per se: both are suspended between subject and object and engage in "transubstantiations" from the organic to the inorganic and the earthly and the divine. In developing his materialism from out of an analysis of literary forms, Tiffany challenges the long-standing norm that regards science as "the sole arbiter in the determination of matter" (75). He wants to pick "the lock that currently bars the literary critic from addressing the problem of material substance" (77).

10 Though, it is more accurate to say that this efficacy belongs less to minerals alone than to the combined activities of a variety of bodies and forces acting as an agentic assemblage.

11 The efficacy of moralism in addressing social problems is overrated.

The antimoralism that is one of the implications of a vital materialism is a dangerous game to play and not one I wish to play out to its logical extreme. I aim not to eliminate the practice of moral judgment but to increase the friction against the moralistic reflex.

12 Romand Coles offers a sustained interpretation of Adorno as an ethical theorist: negative dialectics is a "morality of thinking" that can foster generosity toward others and toward the nonidentical in oneself. Coles argues that Adorno seeks a way to acknowledge and thereby mitigate the violence done by conceptualization and the suffering imposed by the quest to know and control all things (Coles 1997, chap. 2).

13 Adorno (1973) also describes this pain as the "guilt of a life which purely as a fact will strangle other life" (364). Coles (1997) calls it the "ongoing discomfort that solicits our critical efforts" (89). Adorno does not elaborate or defend his claim that the pain of conceptual failure can provoke or motivate an ethical will to redress the pain of social injustice. But surely some defense is needed, for history has shown that even if the pangs of nonidentity engender in the self the idea that "things should be different," this moral awakening does not always result in "social change in practice." In other words, there seems to be a second gap, alongside that between concept and thing, that needs to be addressed: the gap between recognizing the suffering of others and engaging in ameliorative action. Elsewhere I have argued that one source of the energy required is love of the world or enchantment with a world of vital materiality; Adorno sees more ethical potential in suffering and a sense of loss. He "disdained the passage to affirmation," contending that the experience of the "fullness of life" is "inseparable from . . . a desire in which violence and subjugation are inherent. . . . There is no fullness without biceps-flexing" (385, 378). Nonidentity is dark and brooding, and it makes itself known with the least distortion in the form of an unarticulated feeling of resistance, suffering, or pain. From the perspective of the vital materialist, Adorno teeters on the edge of what Thomas Dumm (1999, 169) described as "the overwhelming sense of loss that could swamp us when we approach [the thing's] unknowable vastness."

14 "Preponderance of the object is a thought of which any pretentious philosophy will be suspicious. . . . [Such] protestations . . . seek to drown out the festering suspicion that heteronomy might be mightier

than the autonomy of which Kant . . . taught. . . . Such philosophical subjectivism is the ideological accompaniment of the . . . bourgeois I" (Adorno 1973, 189).

15 The gap between concept and thing can never be closed, and according to Albrecht Wellmer, Adorno believes that this lack of conciliation can be withstood only "in the *name* of an absolute, which, although it is veiled in black, is not nothing. Between the being and the non-being of the absolute there remains an infinitely narrow crack through which a glimmer of light falls upon the world, the light of an absolute which is yet to come into being" (Wellmer 1998, 171; italics added).

16 Thanks to Lars Toender for alerting me to the messianic dimension of Adorno's thinking. One can here note Adorno's admiration for Kant, who Adorno read as having found a way to assign transcendence an important role while making it inaccessible in principle: "What finite beings say about transcendence is the semblance of transcendence; but as Kant well knew, it is a necessary semblance. Hence the incomparable metaphysical relevance of the rescue of semblance, the object of esthetics" (Adorno 1973, 393). For Adorno, "the idea of truth is supreme among the metaphysical ideas, and this is why . . . one who believes in God cannot believe in God, why the possibility represented by the divine name is maintained, rather, by him who does not believe" (401–2). According to Coles, it does not matter to Adorno whether the transcendent realm actually exists; rather, what matters is the "demand . . . placed on thought" by its promise (see Coles 1997, 114).

17 There is, of course, no definitive way to prove either ontological imaginary. Morton Schoolman argues that Adorno's approach, which explicitly leaves *open* the possibility of a divine power of transcendence, is thus preferable to a materialism that seems to close the question. See his *Reason and Horror* (Althusser 2001).

18 In response to Foucault's claim that "perhaps one day, this century will be known as Deleuzean," Deleuze (1995, 88–89) described his own work as naive: "[Foucault] may perhaps have meant that I was the most naive philosopher of our generation. In all of us you find themes like multiplicity, difference, repetition. But I put forward almost raw concepts of these, while others work with more mediations. I've never worried about going beyond metaphysics. . . . I've never

renounced a kind of empiricism. . . . Maybe that's what Foucault meant: I wasn't better than the others, but more naive, producing a kind of *art brut,* so to speak, not the most profound but the most innocent." My thanks to Paul Patton for this reference.

19 There are no supernatural bodies or forces for Lucretius, and if we sometimes seem to have spiritual experiences, that is only because some kinds and collections of bodies exist below the threshold of human sense perception.

20 "Without swerve and encounter, [primordia] would be nothing but *abstract* elements. . . . So much so that we can say that [prior to] . . . *the swerve and the encounter* . . . they led only a phantom existence" (Althusser 2006, 169).

21 Lucretian physics is the basis for his rejection of religion, his presentation of death as a reconfiguration of primordia made necessary by the essential motility of matter, and his ethical advice on how to live well while existing in one's current material configuration.

22 For Adorno (1973, 78), Heidegger, "weary of the subjective jail of cognition," became "convinced that what is transcendent to subjectivity is immediate for subjectivity, without being conceptually stained by subjectivity." But it does not seem to me that Heidegger makes a claim to immediacy (see Heidegger 1967).

23 For Marx, too, naive realism was the philosophy to overcome. He wrote his doctoral dissertation on the "metaphysical materialism" of the Epicureans, and it was against its naive objectivism that Marx would define his own "historical materialism." Historical materialism would not focus on matter but on human-power-laden socioeconomic structures.

24 This is Bill Brown's account of Arjun Appadurai's (1986) project in *The Social Life of Things.* See Brown (2002) for a useful survey of different approaches to the thing.

References

Adorno, Theodor. 1973. *Negative Dialectics.* New York: Continuum.

Althusser, Louis. 2001. *Reason and Horror.* New York: Routledge.

———. 2006. "The Underground Current of the Materialism of the Encounter." In *Philosophy of the Encounter*, trans. François Matheron, Oliver Corpet, and G. M. Goshgarian, 163–207. London: Verso.

Appadurai, Arjun. 1986. *The Social Life of Things*. Cambridge: Cambridge University Press.

Brown, Bill. 2002. "Thing Theory." *Critical Inquiry* 28, no. 1: 1–22.

Brown, Wendy. 1995. *States of Injury*. Princeton, N.J.: Princeton University Press.

Buell, John, and Tom DeLuca. 1996. *Sustainable Democracy: Individuality and the Politics of the Environment*. Thousand Oaks, Calif.: Sage.

Butler, Judith. 1993. *Bodies That Matter*. New York: Routledge.

———. 1998. "Merely Cultural." *New Left Review* 227: 33–44.

Coles, Romand. 1997. *Rethinking Generosity*. Ithaca, N.Y.: Cornell University Press.

De Landa, Manuel. 2000. *A Thousand Years of Nonlinear History*. New York: Zone Books.

———. 2003. *Intensive Science and Virtual Philosophy*. London: Continuum.

Deleuze, Gilles. 1991. *Bergsonism*. New York: Zone Books.

———. 1995. *Negotiations*. Trans. Martin Joughin. New York: Columbia University Press.

De Vries, Hent. 2006. Introduction to *Political Theologies*, ed. Hent de Vries and Lawrence Eugene Sullivan. New York: Fordham University Press.

Dumm, Thomas L. 1999. *A Politics of the Ordinary*. New York: New York University Press.

Edensor, Tim. 2005. "Waste Matter: The Debris of Industrial Ruins and the Disordering of the Material World." *Journal of Material Culture* 10, no. 3: 311–22.

Feher, Michel, ed., with Ramona Naddaff and Nadia Tazi. 1989. *Fragments for a History of the Human Body*. New York: Zone Books.

Ferguson, Kathy. 1991. *The Man Question*. Berkeley: University of California Press.

Frow, John. 2001. "A Pebble, a Camera, a Man." *Critical Inquiry* (Autumn): 270–85.

Gatens, Moira. 1996. *Imaginary Bodies*. New York: Routledge.

Gould, Stephen Jay. 2002. *The Structure of Evolutionary Theory*. Cambridge, Mass.: Harvard University Press.

Heidegger, Martin. 1967. *What Is a Thing?* Trans. W. B. Barton Jr. and Vera Deutsch. New York: Gateway.

Kafka, Franz. 1983. "Cares of a Family Man." In *Complete Stories*, ed. Nahum N. Glatzer, 427–29. New York: Schocken.

Latour, Bruno. 1997. "On Actor-Network Theory: A Few Clarifications." CSI-Paris. http://www.nettime.org/Lists-Archives/nettime-l-9801/msg00019.html.

———. 2004. *The Politics of Nature.* Trans. Catherine Porter. Cambridge, Mass.: Harvard University Press.

Levene, Nancy. 2004. *Spinoza's Revelation.* Cambridge: Cambridge University Press.

Lucretius. 1995. *De Rerum Natura.* In *The Epicurean Philosophers*, ed. John Gaskin, 78–304. New York: Everyman.

Lyotard, Jean-Francois. 1997. *Postmodern Fables.* Trans. Georges van den Abbeele. Minneapolis: University of Minnesota Press.

Margulis, Lynn, and Dorion Sagan. 1995. *What Is Life?* Berkeley: University of California Press.

Mathews, Freya. 2003. *For Love of Matter.* Albany: State University of New York Press.

Melamed, Yitzhak. 2010. "Spinoza's Anti-humanism." In *The Rationalists,* ed. Carlos Fraenkel, Dario Perinetti, and Justin Smith. New York: Springer.

Merleau-Ponty, Maurice. 1981. *The Phenomenology of Perception.* New York: Routledge.

Mitchell, W. J. Thomas. 2005. *What Do Pictures Want? The Lives and Loves of Images.* Chicago: University of Chicago Press.

Pietz, William. 1997. "Death of the Deodand: Accursed Objects and the Money Value of Human Life." *Res* 31: 97–108.

Rahman, Momin, and Anne Witz. 2002. "What Really Matters? The Elusive Quality of the Material in Feminist Thought." Paper presented to the annual congress of the Canadian Sociology andAnthropology Association, Toronto.

Rorty, Richard. 1995. *Rorty and Pragmatism: The Philosopher Responds to His Critics.* Ed. Herman J. Saatkamp Jr. Nashville, Tenn.: Vanderbilt University Press.

Sharp, Hasana. 2007. "The Force of Ideas in Spinoza." *Political Theory* 35, no. 6: 732–55.

Spinoza, Baruch. 1982. *The Ethics with the Treatise on the Emendation of the Intellect and Selected Letters.* Trans. Samuel Shirley. Indianapolis, Ind.: Hackett.

———. 1995. *Spinoza: The Letters.* Trans. Samuel Shirley. Indianapolis, Ind.: Hackett.

Sullivan, Robert. 1998. *The Meadowlands: Wilderness Adventures on the*

Edge of the City. New York: Doubleday Anchor Books.
Thoreau, Henry. 1949. *The Journal of Henry David Thoreau.* Vol. 4. Ed. Bradford Torrey and Francis H. Allen. New York: Houghton Mifflin.
———. 1973. *The Writings of Henry David Thoreau: Walden.* Ed. J. Lyndon Shanley. Princeton, N.J.: Princeton University Press.
Tiffany, Daniel. 2001. "Lyric Substance: On Riddles, Materialism, and Poetic Obscurity." *Critical Inquiry* 28: 72–98.
Wellmer, Albrecht. 1998. *Endgames: The Irreconcilable Nature of Modernity,* trans. David Midgley. Cambridge, Mass.: MIT Press.

3 Materiality, Experience, and Surveillance

WILLIAM E. CONNOLLY

Nature, Culture, Immanence

I SEEK TO COME TO TERMS with the materiality of perception by placing Merleau-Ponty, Michel Foucault, and Gilles Deleuze into conversations with each other and neuroscience. Such a conversation has been obstructed by the judgment that Merleau-Ponty is a phenomenologist whereas the latter two are opposed to phenomenology. My sense, however, is that there is a phenomenological moment in both Foucault and Deleuze. Moreover, the conception of the subject they criticize is one from which Merleau-Ponty progressively departed. He also moved toward a conception of nonhuman nature that, he thought, was needed to redeem themes in the *Phenomenology of Perception.* This double movement—revising the idea of the subject and articulating a conception of nature compatible with it—draws Merleau-Ponty closer to what I will call a philosophy of immanence. Whether that migration was completed or punctuated by a moment of transcendence is a question I will not answer here.

By *immanence* I mean a philosophy of becoming in which the universe is not dependent on a higher power. It is reducible neither to mechanistic materialism, dualism, theo-teleology, nor the absent God of minimal theology. It concurs with the last three philosophies that there is more to reality than actuality. But that "more" is not given by a robust or minimal God. We bear no debts or primordial guilt for being, even if there are features of the human condition that tempt many to act as if we do;[1] rather, there are uncertain exchanges between stabilized formations and mobile forces that subsist within and below them. Biological evolution, the evolution of the universe, radical changes in politics, and the significant conversion experiences of individuals attest to the periodic amplification of such circuits of exchange.

Gilles Deleuze and Félix Guattari state the idea this way. First, they challenge the idea of transcendence lodged "in the mind of a god, or in the unconscious of life, of the soul, or of language . . . always inferred" (Deleuze and Guattari 1987, 266). Second, they affirm historically shifting "relations of movement and rest, speed and slowness between unformed elements, or at last between elements that are relatively unformed, molecules and particles of all kinds" (266). Such a philosophy of "movement and rest" does not imply that everything is always in flux, though its detractors often reduce it to that view.[2] It means that though any species, thing, system, or civilization may last for a long time, nothing lasts forever. Each force field (set in the chrono-time appropriate to it) oscillates between periods of relative arrest and those of heightened imbalance and change, followed again by new stabilizations. Neither long cycles of repetition, nor linear causality, nor an intrinsic purpose exist in being, but, as the Nobel Prize–winning chemist Ilya Prigogine (2003, 65) puts it, "our universe is far from equilibrium, nonlinear and full of irreversible processes."

There is no denial that we humans—while often differing from one another—*judge* the new outcomes to which we are exposed or have helped to usher into being. What is denied is that the judgments express an eternal law or bring us into attunement with an intrinsic purpose of being. For immanent materialists deny there is such a law or intrinsic purpose. We anchor our ethics elsewhere and in a different way. Immanent materialism is defined by contrast to mechanistic materialism, too. Many causal relations are not susceptible to either efficient or mechanical modes of analysis. There are efficient causes, as when, to take a classic example, one billiard ball moves another in a specific direction. But *emergent causality*—the dicey process by which new entities and processes periodically surge into being—is irreducible to efficient causality. It is a mode in which new forces can trigger novel patterns of *self-organization* in a thing, species, system, or being, sometimes allowing something new to emerge from the swirl back and forth between them: a new species, state of the universe, weather system, ecological balance, or political formation.

Merleau-Ponty traveled from his early work on perception to an image that draws humanity closer to the rest of nature than dominant philosophies of the past had proposed. A certain pressure to pursue that journey was always there: a layered theory of human embodiment faces pressure to identify selective *affinities* between the capacities of humans and other living beings and physical systems.

Consider some statements from *Nature,* a collection of lectures given by Merleau-Ponty just before his untimely death:

> Thus, for instance, the Nature in us must have some relation to Nature outside of us; moreover, Nature outside of us must be unveiled to us by the Nature that we are. . . . We are part of some Nature, and reciprocally, it is from ourselves that living beings and even space speak to us. (Merleau-Ponty 2003, 206)

Here Merleau-Ponty solicits affinities between human and nonhuman nature. Does he also suggest that once preliminary affinities have been disclosed, it is possible to organize experimental investigations to uncover dimensions of human and nonhuman nature previously outside the range of that experience? And that these findings might then be folded into an enlarged experience of ourselves and the world?[3] If so, when the neuroscientist V. S. Ramachandran, using magnetic imaging and other technologies of observation, exposes body–brain processes in the production of phantom pain exceeding those assumed in Merleau-Ponty's experiential account of it, those findings could be folded into the latter's account along with the techniques Ramachandran invented to relieve such pain[4] (Ramachandran 1998). Here *experimental* and *experiential* perspectives circulate back and forth, with each sometimes triggering a surprising change in the other. Consider another formulation:

> All these ideas (vitalism, entelechy) suppose preformation, yet modern embryology defines the thesis of epigenesis. . . . The future must not be contained in the present. . . . It would be arbitrary to

> understand this history as the epiphenomenon of a mechanical causality. Mechanistic thinking rests upon a causality which traverses and never stops in something. (Merleau-Ponty 2003, 152)

"The future must not be contained in the present." Just as the future of human culture is not sufficiently determined by efficient causes from the past, in nonhuman nature, too—when the chrono-periods identified are appropriate to the field in question—the future is not sufficiently contained in the present. Now mechanical causality, vitalism, and entelechy, on Merleau-Ponty's reading of them at least, bite the dust together.

But if the future is not sufficiently contained in the present, what enables change over short and long periods? Here Merleau-Ponty approaches an orientation now familiar in the work of scientists such as Ilya Prigogine in chemistry, Brian Goodwin and Lynn Margulis in biology, Antonio Damasio and Ramachandran in neuroscience, and Stephen Gould in evolutionary biology:[5]

> The outlines of the organism in the embryo constitute a factor of imbalance. It is not because humans consider them as outlines that they are such but because they break the current balance and fix the conditions for a future balance. (Merleau-Ponty 2003, 156)

The "imbalance" noted by Merleau-Ponty is close to what Gilles Deleuze calls the "asymmetry of nature," an energized asymmetry that periodically sets the stage, when other conditions are in place, for old formations to disintegrate and new ones to surge into being. It bears a family resemblance to Prigogine's account of systems that enter a period of "disequilibrium" and to the behavior "on the edge of chaos" that Brian Goodwin studies when a species either evolves into a new, unpredictable one or faces extinction. Merleau-Ponty, in alliance with these thinkers, does not shift from a mechanical conception of natural order to a world of chaos. He suggests that in each object domain, periods of imbalance alternate with those of new and imperfect stabilizations. I take these formulations to support the adventure pursued here.

The Complexity of Perception

Visual perception involves a complex mixing—during the half-second delay between the reception of sensory experience and the formation of an image—of language, affect, feeling, touch, and anticipation.[6] This mixing is set in the memory-infused life of human beings whose experience is conditioned by the previous discipline of the chemical-electrical *network* in which perception is set and the characteristic *shape* of human embodiment and motility. Human mobility is enabled by our two-leggedness and the position of the head at the top of the body, with two eyes pointed forward. This mode of embodiment, for instance, encourages the production of widespread analogies between a future "in front of us" and the past "behind us." Most important, the act of perception is permeated by implicit reference to the position and mood of one's own body in relation to the phenomenal field (Merleau-Ponty 2003, 100). Experience is grasped, says Merleau-Ponty (1962, 52), "first in its meaning for us, for that heavy mass which is our body, whence it comes about that it always involves reference to the body." My "body appears to me as an attitude directed towards a certain existing or possible task. And indeed its spatiality is not a *spatiality of position but a spatiality of situation*" (Merleau-Ponty 1962, 100).

We also need to come to terms with how perception is *intersensory*, never fully divisible into separate sense experiences.[7] For example, visual experience is saturated with the tactile history of the experiencing agent. The tactile and the visual are interwoven in that my history of touching objects similar to the one in question is woven into my current vision of it. A poignant example of this is offered by Laura Marks as she elucidates a film scene in which the composition of voice and the grainy visual image convey the daughter's tactile memory of her deceased mother's skin (Marks 2003).

Similarly, language and sense experience are neither entirely separate nor reducible to one another. They are imbricated in a way that allows each to exceed the other in experience: "the sense being held within the word, and the word being the external existence of the sense" (Merleau-Ponty 1962, 183).[8]

Continuing down this path, Merleau-Ponty indicates how the color of an object triggers an affective charge. People with specific motor disturbances take jerky movements if the color field is blue and more smooth ones if it is red or yellow. And in so-called normal subjects, too, the visual field of color is interwoven with an experience of warmth or coldness that precedes and infuses specific awareness of it, depending on whether the field is red or blue (Merleau-Ponty 1962, 209, 211). This field of interinvolvement, in turn, flows into that between color and sound, in which specific types of sound infect the experience of color, intensifying or dampening it (228). Words participate in this process, too, as when the "word 'hard' produces a stiffening of the back or neck." Even "before becoming the indication of a concept the word is first an event which grips my body, and this grip circumscribes the area of significance to which it has reference" (235). The "before" in this sentence does not refer to an uncultured body but to a preliminary tendency in encultured beings. To put the point another way, the imbrications between embodiment, language, disposition, perception, and mood are always in operation. A philosophy of language that ignores these essential connections may appear precise and rigorous, but it does so by missing circuits of interinvolvement through which perception is organized.

These preliminary experiences vary across individuals and cultures, and those variations are important to an appreciation of cultural diversity. The key point, however, is that some series of interinvolvements is always encoded into the preliminary character of experience, flowing into the tone and color of perception. Phenomenologists, Buddhist monks, corporate advertisers, cultural anthropologists, neuroscientists, TV dramatists, Catholic priests, filmmakers, and evangelical preachers are attuned to such memory-soaked patterns of interinvolvement. Too many social scientists, analytic philosophers, rational choice theorists, deliberative democrats, and "intellectualists" of various sorts are less so. An intellectualist, to Merleau-Ponty, is one who overstates the autonomy of conceptual life, the independence of vision, the self-sufficiency of reason, the power of pure deliberation, and/or the self-sufficiency of argument.

Perception not only has multiple layers of intersensory *memory* folded into it, it is suffused with *anticipation.* This does not mean merely that you anticipate a result and then test it against the effect of experience. It means that perception expresses a set of anticipatory expectations that help to constitute what it actually becomes. The case of the word *hardness* already suggests this. A more recent experiment by neuroscientists dramatizes the point. The body–brain patterns of the respondents were observed through various imaging techniques, and the subjects were asked to follow a series of pictures moving from left to right. The images at first glance look the same, but on closer inspection, your experience shifts abruptly from that of the bare head of a man to the nude body of a woman as you proceed down the line of images. People vary in terms of the point at which the gestalt switch occurs. More compellingly, when asked to view the series a second time from right to left, almost everyone identifies the shift from the nude woman to the man's face further down the trail than she had in moving from left to right. The authors contend that the body–brain processes catalyzed by this series engender dicey transitions between two embodied attractors. The first attractor retains its hold as long as possible; the second, triggered as you move from right to left, is retained until pressed to give way to another. The suddenness of shift in experience correlates with dramatic shifts in observable body and brain patterns:

> By placing electrodes on the appropriate muscles to measure their electromagnetic activity, Kelso could clearly measure the sudden shift from one pattern to another. The underlying idea in Kelso's studies was that the brain is a self-organizing, pattern forming system that operates close to instability points, thereby allowing it to switch flexibly and spontaneously from one coherent state to another. (Sole and Goodwin 2000, 142–43)

The "imbalance" that Merleau-Ponty identifies in embryos also operates in the perception of mobile human beings who must respond to rapidly shifting contexts.[9] Perception, to be flexible, is organized through multiple points of "instability," through which one set of memory-infused attractors gives way to another when the pressure of the encounter becomes intense enough. Each attractor helps to structure the actuality of perception.

Perception could not function without a rich history of interinvolvements between embodiment, movement, body image, touch, sight, smell, language, affect, and color. The anticipatory structure of perception enables it to carry out its functions in the rapidly changing contexts of everyday life; it also opens it to subliminal influence by mystics, priests, lovers, politicians, parents, military leaders, filmmakers, teachers, talk show hosts, and TV advertisers.

Another way of putting the point is to say that the actuality of perception is "normative," where that word now means the application of a culturally organized attractor to a situation roughly responsive to it. A visual percept, for instance, contains the norm of a well-rounded object, compensating for the limitations of the particular position from which it starts. As Merleau-Ponty puts it, "the unity of either the subject or the object is not a real unity, *but a presumptive unity on the horizon of experience.* We must rediscover, as anterior to the ideas of subject and object . . . that primordial layer at which both things and ideas come into being" (Merleau-Ponty 1962, 219; italics added). The import of this presumptive unity becomes more clear through the discussions of depth and discipline.

Visibility and Depth

Merleau-Ponty concludes that we make a singular contribution to the experience of spatial depth, even though, as Diana Coole (2000, 132) says, "the depth and perspective that permit visual clarity belong to neither seer nor seen [alone], but unfold where they meet."[10] The experience of depth, you might say, incorporates different possible perspectives on the object *into* the angle of vision from which it is now engaged. The experience is ubiquitous. If you draw a Necker cube on a flat piece of paper, depth will immediately be projected into it. On viewing the image for a few seconds, the image becomes inverted so that a figure in which depth had moved from left to right now flips in the other direction. On learning how to produce the flips—by focusing your eye first on the bottom right angle and then on the top left angle—it becomes clear how difficult it is to purge experience of depth. The short interval between the switch of gaze and the flip of the angle also testifies to the half-second delay between the reception of sensory experience and cultural participation in the organization of perception. It teaches us that perception must be disciplined and draws attention to the fugitive interval during which that organization occurs. René Magritte dramatizes the vertigo that arises when anticipation of depth is stymied in *The Blank Signature*. A woman is riding a horse in the woods. But the lines of visibility and invisibility in the painting confound those we anticipate. The horse's back left leg curls behind a tree trunk in front of its torso, and just where the scenic background should slide behind the horse, it appears in front of one part of its torso. Now you have a strangely familiar scene in which it is impossible to redeem the depth experience solicits. The painting dramatizes how the visible, set against a field of the invisible, ordinarily helps to set the place, import, and depth of an image. The power of depth to insist is further emphasized when you discern how the anticipation of it is realized to the immediate left and right of the woman on the horse. But what enables the experience of depth when perception is unencumbered by such contrivances?

Perception depends on projection into experience of multiple perspectives you do not now have. This automatic projection into experience also makes it seem that objects see you as you see them. Merleau-Ponty (1968, 13; italics added) puts it this way: in this "strange adhesion of the seer and the visible . . . *I feel myself looked at by the things*, my activity is equally a passivity."[11] To have the experience of depth is to feel things looking at you, to feel yourself as object. This self-awareness is usually subliminal, but it becomes more apparent when you shift from the process of action-oriented perception to dwell in experience itself. The result is uncanny: to see is to experience oneself as an object of visibility, not simply in that you realize someone *could* look at you because you are composed of opaque materiality but also because the very structure of vision incorporates into itself the projection of what it would be like to be seen from a variety of angles. This experience codifies, in the anticipatory structure of perception, potential angles of vision on you and what it would be like to touch, hold, or move the object from different angles. The codification of operational angles of possible action and the background sense of being seen combine to produce depth.

That codification, however, cannot be reduced to the sum of all angles, to a view from *nowhere*. It cannot because each potential angle of vision fades into a diffuse background against which it is set. The codification, then, is closer to a view from *everywhere*, a view projected as a norm into an experience that depends on implicit reference to it. In an essay on Merleau-Ponty, Sean Dorrance Kelly pulls these themes of anticipation and perspective together. First, the experience of a particular light or color is normative in the sense that "each presentation of the color in a given lighting . . . makes an implicit reference to a more completely presented *real* color . . . if the lighting context were changed in the direction of the norm. This real color, implicitly referred to . . . is the constant color I see the color to be." Second, "the view from everywhere" built into the experience of depth is not a view you could ever actually have, separate from these memory soaked projections, because there is no potential perspective that could add up the angles and backgrounds appropriate to all perspectives. Backgrounds are not additive in this way. The experience of depth is rather "a view . . . from which my own perspective is felt to deviate" (Kelly 2005, 85, 92). The perception of depth anticipates a perspective from which my actual angle of vision is felt to deviate. Perception thus closes into itself *as* actuality, a norm it cannot in fact instantiate. Perception *is* anticipatory and normative. The only thing Kelly omits is how the perception of depth is also one in which "I feel myself looked at by things," in which my activity of perception "is equally a passivity." That theme has consequences for contemporary politics.

Perception and Discipline

It might still seem that the gap between Michel Foucault and Merleau-Ponty remains too large to enable either to illuminate the other. Did not the early Foucault argue that because of the opacity of "life, labor and language," the structure of experience cannot provide a solid base from which to redeem a theory of the subject? Did he not say that the transcendental arguments that phenomenologists seek—whereby you first locate something indubitable in experience and then show what conception of the subject is necessarily presupposed by that experience—cannot be stabilized when the "doubles" of life, labor, and language fade into obscurity? Yes. But those strictures may be more applicable to Husserl than to Merleau-Ponty, particularly in the latter's later work.

Foucault speaks of "discipline" as a political anatomy of detail that molds the posture, demeanor, and sensibilities of the constituencies subjected to it, "in which power relations have an immediate hold on [the body]; they invest it, mark it, train it, torture it, force it to carry out tasks, to perform ceremonies, to enact signs" (Foucault 1977, 25). We note already a difference in rhythm between the sentences of Foucault and those of Merleau-Ponty. Merleau-Ponty's sentences convey an implicit sense of belonging to the world, whereas Foucault's often identify or mobilize elements of resistance and disaffection circulating within modern modalities of experience. The initial connection between these two thinkers across difference is that both see how perception requires a prior *disciplining* of the senses in which a rich history of interinvolvement sets the stage for experience. The critical relation between corporeocultural discipline and the shape of experience is emphasized by the fact that adults who have the neural machinery of vision repaired after having been blind from birth remain operationally blind unless and until a new history of interinvolvements between movement, touch, and object manipulation is synthesized into the synapses of the visual system. Only about 10 percent of the synaptic connections for vision are wired in at birth. The rest emerge from the interplay between body–brain pluripotentiality and the history of intersensory experience.[12]

Let us return to Merleau-Ponty's finding that to perceive depth is implicitly to feel yourself as an object of vision. In a disciplinary society, this implicit sense morphs into a more intensive experience of being an actual or potential object of *surveillance* in a national security state. That latter experience was amplified in the United States after the al-Qaeda attack of 9/11, the event in which Osama bin Laden invited George W. Bush to organize the world through the prism of security against a pervasive, nonstate enemy, and the cowboy eagerly accepted the invitation. The indubitable experience of self-visibility now swells into that of being an object of surveillance. Everyday awareness of that possibility recoils back on the shape and emotional tone of experience. Traffic cameras, airport screening devices, the circulation of social security numbers, credit profiles, medical records, electric identification bracelets, telephone caller ID, product surveys, National Security Agency sweeps, telephone records, license plates, Internet use profiles, Internal Revenue Service audits, driver's licenses, police phone calls for "contributions," credit card numbers, DNA records, fingerprints, smellprints, eyeprints, promotion and hiring profiles, drug tests, street and building surveillance cameras: several of these are used on, in, or at work, voter solicitation, the school, the street, job interviews, police scrutiny, prison observation, political paybacks, racial profiling, e-mail solicitations, church judgments, divorce proceedings, and the publication of sexual proclivities. As such devices proliferate, the experience of *potential* observability becomes an active element in experience:[13]

> A whole problematic then develops: that of an architecture that is no longer built simply to be seen . . . or to observe the external space . . . but to permit an internal, articulated and detailed control—to render visible those who are inside it . . . an architecture that would operate to transform individuals: to act on those it shelters, to provide a hold on their conduct, to carry the effects of power right to them, to make it possible to know them, to alter them. (Foucault 1977, 172)

True, Foucault's description of disciplinary society does not deal adequately with differences in age, class, and race. There is today an urban underclass that is subjected to general strategies of urban *containment* and impersonal modes of *surveillance* in stores, streets, public facilities, reform schools, prisons, and schools. There is also a suburban, upper-middle, career-oriented class enmeshed in detailed disciplines in several domains, anticipating the day it rises above them. And there are several other subject positions, too, including those who rise more or less above generalized surveillance.

Watch out. Are you a war dissenter? Gay? Interested in drugs? An atheist who talks about it? A critic of the war on terrorism, drug policies, or government corruption? Sexually active? Be careful. You may want a new job someday or to protect yourself against this or that charge. Protect yourself now in anticipation of uncertain possibilities in the future. Discipline yourself in response to future threats. In advanced capitalism, where the affluent organize life around the prospect of a long career, many others look for jobs without security or benefits, and others yet find themselves stuck in illegal, informal, and underground economies, the implicit message of the surveillance society is to remain unobtrusive and politically quiescent by appearing more devout, regular, and patriotic than the next guy. The implicit sense of belonging to the world that Merleau-Ponty found folded into the fiber of experience now begins to ripple and scatter.

Neither Foucault nor Merleau-Ponty, understandably, was as alert to the electronic media as we must be today. This ubiquitous force flows into the circuits of discipline, perception, self-awareness, and conduct. It is not enough to survey the pattern of media ownership. It is equally pertinent to examine the methods through which it becomes insinuated into the shape and tone of perception.

Here I note one dimension of a larger topic. To decode electoral campaigns, it is useful to see how media advertising works. According to Robert Heath, a successful ad executive and follower of recent work in neuroscience, the most effective product ads target viewers who are distracted from them. The ad solicits "implicit learning" below the level of refined intellectual attention. It plants "triggers" that insinuate a mood and/or association into perception, called into action the next time the product is seen, mentioned, smelled, heard, or touched. Implicit learning is key because, unlike the refined intellectual activity into which it flows, "it is on all the time." It is "automatic, almost inexhaustible, in its capacity and more durable" in retention (Heath 2005, 67).[14]

The link to Foucault and Merleau-Ponty is that they, too, attend to the preconscious, affective dimensions of discipline and experience without focusing on the media. Today, programs such as the *Hannity-Colmes Report, Crossfire,* and *The O'Reilly Factor* infiltrate the tonalities of political perception. As viewers focus on points made by guests and hosts, the program is laced with interruptions, talking-over, sharp accusations, and yelling. The endless reiteration of those intensities secretes a simple standard of objectivity as the gold standard of perception while insinuating the corollary suspicion that no one actually measures up to it. As a result, resentment and cynicism now become coded into the very color of perception. The cumulative result of the process itself favors a neoconservative agenda. For cynics typically ridicule the legacy of big government in employment, services, and welfare while yearning for a figure to reassert the unquestionable authority of the "nation." A cynic is an authoritarian who rejects the current regime of authority. Cynical realists experience the fragility and uncertainty that help to constitute perception. But they join that experience with an overweaning demand for authority, and they accuse everyone else of failing to conform to the model of simple objectivity they claim to meet. Justification of this model is not sustained by showing how they meet it but by repeated accusations that others regularly fail to do so.

Cynical realism is one response to the complexity of perception. Another, in a world of surveillance, is self-depoliticization. You avert your gaze from disturbing events to curtail dangerous temptations to action. The goal is to avoid close attention or intimidation in the venues of work, family, school, church, electoral politics, and neighborhood life. But of course, such a retreat can also amplify a feeling of resentment against the organization of life itself, opening up some of these same constituencies for recruitment by the forces of *ressentiment*. Such responses can be mixed in several ways. What is undeniable is that the circuits between discipline, media, layered memories, and self-awareness find expression in the color of perception itself. Power is coded into perception.

The Micropolitics of Perception

Sensory interinvolvement, disciplinary processes, detailed modes of surveillance, media practices of infiltration, congealed attractors, affective dispositions, self-regulation in response to future susceptibility—these elements participate in perpetual circuits of exchange, feedback, and reentry, with each loop folding another variation and degree into its predecessor. The imbrications are so close that it is impossible to sort out each. The circuits fold, bend, and blend into each other, inflecting the shape of political experience. Even as they are ubiquitous, however, there are numerous points of dissonance, variation, hesitation, and disturbance in them. These interruptions provide potential triggers to the pursuit of other possibilities. A past replete with religious ritual clashes with an alternative representation of God in a film, church, or school; an emergent practice of heterodox sexuality encourages you to question established habits in other domains; the interruption of a heretofore smooth career path solicits doubts previously submerged in habits of anticipation; a trip abroad exposes you to disturbing news items and attitudes seldom allowed expression in your own country; neurotherapy fosters a modest shift in your sensibility; a stock market crash disrupts assumptions about the future; a new religious experience shakes you; a terrorist attack folds an implacable desire for revenge into you; a devastating natural event shakes your faith in providence. The anticipatory habits of perception are not self-contained; rather, dominant tendencies periodically bump into minor dispositions, submerged tendencies, and uncertain incipiencies. The instability of the attractors and conjunctions that make perception possible also renders it a ubiquitous medium of power and politics. What might be done today to open habits of the anticipation of more constituencies during a time when media politics diverts attention from the most urgent dilemmas of the day?

Television could be a site on which to run such experiments. A few dramas do so. I would place *Six Feet Under* on that list, as it disrupts a conventional habit of perception and occasionally works to recast it. But the closer a program is to a "news program" or a "talk show," the more it either enacts virulent partisanship, adopts the hackneyed voice of simple objectivity, or both. What is needed are subtle media experiments, news and talk shows that expose and address the complexity of experience in a media-saturated society. *The Daily Show* and *The Colbert Report* take a couple of steps in the right direction, calling into question the voice of simple objectivity through exaggeration and satirization of it. But much more is needed.

Mark Hansen, in *New Philosophy for New Media*, pursues this issue. In chapter 6, he reviews *Skulls*, an exhibit presented by Robert Lazzarini at the Whitney Museum in 2000. Lazzarini's sculptures are uncanny. They seem like skulls, but you soon find that however you tilt your head or change your position, it is impossible to vindicate the anticipation of them. Lazzarini has in fact laser scanned an actual human skull, reformatted it into several images, and constructed a few statues from the reformatted images. Now no three-dimensional image can be brought into alignment with the anticipation triggered by its appearance. "At each effort to align your point of view with the perspective of one of these weird sculptural objects, you experience a gradually mounting feeling of incredible strangeness. *It is as though these skulls refuse to return your gaze*" (Hansen 2004, 198).

The sense of being seen by the objects you see is shattered by deformed images that refuse to support that sense. You now feel "the space around you begin to ripple, to bubble, to infold, as if it were becoming unstuck from the fixed coordinates of its three dimensional extension" (Hansen 2004, 198). *Skulls,* when joined to Merleau-Ponty's phenomenology of perception, heightens awareness of the fugitive role we play in perception by making it impossible to find an attractor to which it corresponds. These sculptures also dramatize the role that *affect* plays in perception, as they jolt the tacit feeling of belonging to the world that Merleau-Ponty imports into the depth grammar of experience. The implicit sense of belonging to the world is transfigured into a feeling of vertigo. Do such experiments dramatize a condition that is already lurking within experience in a new world marked by the acceleration of tempo and the exacerbation of surveillance? Or do they show how that sense is still secure when not disrupted by disturbing experiences? That is a key question. At a minimum, in conjunction with the work of Merleau-Ponty and Foucault, they sharpen awareness of the memory-soaked imbrications between affect, tactility, and vision in the process of perception. You now call into question simple models of vision and better appreciate how a disciplinary society inflects affect-imbued perception. You might even become attracted to experimental strategies to deepen visceral attachment to the complexity of existence itself during a time when the automatic sense of belonging is often stretched and disrupted.

None of the preceding responses is automatic. A door merely opens. Walking through it, more of us may face the complexity of experience and resist the drive to existential resentment that often surges up when it is confronted. How could such a temper be cultivated?

As a preliminary, consider some processes and conditions that disrupt the implicit sense of belonging to the world. They include the acceleration of speed in many domains of life, including military deployment, global communication systems, air travel, tourism, population migrations, fashion, economic transactions, and cultural exchanges; a flood of popular films that complicate visual experience and call the linear image of time into question; publicity about new discoveries in neuroscience, which include attention to that half-second delay between sensory reception and the organization of perception; greater awareness of work in several domains of science that transduct the Newtonian model of linear cause into the ontological uncertainty of emergent causality; scientific speculations that extend the theory of biological evolution to the unfolding of the universe itself; increased media attention to events that periodically shock habitual assumptions coded into perception; media attention to the devastation occasioned here or there by earthquakes, hurricanes, volcanic eruptions, and tsunamis; and a vague but urgent sense that the world's fragile ecological balance is careening into radical imbalance.

The signs that these disruptive experiences have taken a toll are also multiple. They include, on the aggressive/defensive side, the extreme levels of violence and superhuman heroism in action films, as they strive to redeem the simple model of objectivism under unfavorable circumstances; the intensification of accusatory voices in the media in conjunction with righteous self-assertions of objectivism; new intensities of apocalyptic prophecy in several religious movements; the virulence of electoral campaigns; and the desire for abstract revenge finding expression in preemptive wars, state regimes of torture, massacres, collective rapes, and the like. Do those actions express covert drives to take revenge against the very terms of modern existence itself?

The obverse side of those responses is discernible in other pressures and constituencies. Today more people are less convinced of the simple model of perception as they seek to consolidate attachment to a world populated by sensory interinvolvements, attractors, the complexity of duration, time as becoming, and an uncertain future. Take, for instance, the receptive responses to minor films such as *Far from Heaven, I Heart Huckabees, Time Code, Blow-up, The Eternal Sunshine of the Spotless Mind, Memento, Waking Life, Run Lola Run,* and *Mulholland Drive.*

These films focus on the role of duration in perception, scramble old habits in this way or that, highlight sensory interinvolvement, challenge simple objectivism, and call the self-certainty of the linear image of time into question. Some take another turn as well. Going through and beyond the anxiety fomented by *Skulls*, they encourage an awakening that is most apt to emerge *after* such anxieties have been tapped. To take one example, *I Heart Huckabees* hints at multiple modes of human and nonhuman agency in a pluralistic universe as it embraces practices it first subjects to mockery, including "existential detectives," ecological movements, media advertisements on their behalf, and micropolitics to increase our attachment to the new world. The humor is folded into the attachments. It is practiced on behalf of human self-modesty in a world in which temporality may be open, human efficacy is fragile, perception is complex, and the myths of providence, intellectualism, and world mastery are open to challenge.[15]

This is the juncture at which the experimental engagements with film by Gilles Deleuze in *Cinema 2* can be placed into engagement with Merleau-Ponty on the complexity of perception and Foucault on modern modes of surveillance. The stage is set for the theme of chapter 7 in *Cinema 2* by discussions of flashbacks that expose moments of bifurcation in experience, comedic figures who enact exquisite sensitivity to movements of the world, irrational cuts that scramble the action image, crystals of time that enact the complexity of duration, and engagements with "powers of the false" that open up old patterns of incipience during the organization of perception. The suggestion is that most of us have already been infected by such experiences in daily life and by films that dramatize and extend them. Such dramatizations can, of course, amplify existential resentment, magnetize drives to reassert the simple model of objectivity, or encourage a retreat from public engagement. But Deleuze challenges all these responses. He encourages tactics to deepen attachment to the complexity of this world itself so as to challenge bellicose mastery, passive skepticism, and authoritarian cynicism at their nodal points of inception.

By "this world," he does not mean the established distribution of power and political priorities. He means affirmation of the larger compass of being in which humans are set as opposed to existential resentment of it or resignation about it. He realizes that people bring different interpretations of "the larger compass of being," but he also contends that the quality of existential temper insinuated into those different beliefs makes a difference to the shape and direction of political life. As he puts it, "whether we are Christians or atheists, in our universal schizophrenia, we need reasons to believe in this world" (Deleuze 1989, 172).

In his usage, the term *belief* cuts deep into incipient dispositions that infect the color of perception. This is the zone that prophets tap, and one the media engages too through the interplay of rhythm, image, music, and sound. Belief touches, for instance, the tightening of the gut, coldness of the skin, contraction of the pupils, and hunching of the back that occur when a judgment or faith in which you are deeply invested is contested or ridiculed. It also touches those feelings of abundance and joy that emerge periodically when we sense the surplus of life over the structure of our identities. That is the surplus Deleuze seeks to mobilize.

It may be important to follow Deleuze's lead, in part because the mode of belonging embraced by Merleau-Ponty has been shaken by the acceleration of pace in many zones of culture and the pervasive role of the media in everyday life.[16]

Wider negotiation of attachment to the most fundamental terms of modern existence would not sanction existing injustices, nor would it *suffice* to spawn the critical politics needed today (though some intellectualists will project both assumptions into this essay). Such energies, rather, must be cultivated more widely and inserted into larger circuits of political action. For we no longer inhabit a world where a sense of belonging is securely installed in the infrastructure of experience, if we ever did. Nor is it likely that a single religious faith can be drawn on to repair the deficit, at least without introducing massive repression into a world where minorities of many types now inhabit the same territorial spaces. The issue is fundamental.

Let us tarry on the question of existential ethos a bit. My experience is that many on the Left who point correctly to the insufficiency of such awakenings move quickly from that point to assert its irrelevance or to announce its foolishness. They do not want to seem soft or feminine. Those are the judgments I contest. Work on the infrastructure of perception is crucial in close conjunction with other modes of politics. To ignore the first dimension is to forfeit too much to the radical Right as it works to fold dispositions to fear, anger, and revenge into the very texture of experience. It first promotes the modes of incipience it seeks and then harvests the dispositions it has fomented. So media and film experiments that incite attachment to a world in which things move faster than heretofore are not to be demeaned. They do their part. They provide triggers and catalysts to a more radical, pluralist, and egalitarian politics. These catalysts are comparable, in their way, to the televangelism of right-wing preachers and Fox talking heads who incite a will to revenge as they discount our responsibility to the future of the earth. The difference is that the preachers inject resentment of this world into the circuits of perception, whereas the need is to solicit a more profound attachment to the future of the earth.

Right-wing politics is paradoxical. It resists academic studies of how perception works but is highly attuned to the operational politics of perception. That combination forms a powerful political formula as it poses a threat to a democratic future. The habits of intellectualism, still haunting the left, take a toll on efforts to forge a counterpolitical formula, one that infuses care for this world into militant critiques of current priorities and carries both into a positive political agenda active on several sites.

Notes

1 You could speak, as Merleau-Ponty occasionally does, of transcendence without the Transcendent. But such a formulation tends to blur the contestation between alternative faith and philosophies that needs to be kept alive.

2 In the introduction to *Problems and Methods in the Study of Politics*, Shapiro, Smith, and Masoud (2004, 11) reported me to say "that the world is in a state of constant and unpredictable flux." That signifies to me that awareness of one side of my position has been blocked by the shock of meeting the other.

3 The formulation in fact suggests the doctrine of parallelism introduced by Spinoza in the seventeenth century. For a fine study in neuroscience that draws on both Spinoza's philosophy of parallelism

and his idea that affect always accompanies perception, belief, and thinking, see Damasio (2003).

4 Some implications of this research for cultural theory are explored by Connolly; see *Neuropolitics: Thinking, Culture, Speed*, chapter 1.

5 See, besides the preceding references to Damasio, Prigogine, and Ramachandran, Brian Goodwin (1994). In *The Structure of Evolutionary Theory*, Stephen Jay Gould (2002) emphasizes how close his revision of Darwinian theory is to the notion of genealogy developed by Nietzsche.

6 The phrase "the half-second delay" comes out of work in neuroscience pioneered by Benjamin Libet. Merleau-Ponty was certainly aware of a time lag, however. An excellent discussion of the delay and its significance can be found in Brian Massumi (2002).

7 This theme is increasingly accepted in neuroscience today. See "Why You Have at Least 21 Senses," in *New Scientist* (2005). The authors of this article agree, too, that the senses are interinvolved.

8 Such a pattern of interinvolvement will only seem impossible to those who are captured by the analytic–synthetic dichotomy, in which every connection is reducible either to a definitional or an empirical (causal) relation. Once you break that dichotomy, you can come to terms with the series of memory-infused interinvolvements through which perception is organized. You are also able to consider models of causality that transcend efficient causality.

9 In fact, Henri Bergson is better than Merleau-Ponty at focusing attention on the role that the imperative to make perceptual judgments rapidly as you run through the numerous encounters of everyday life plays in creating the subtractions and simplifications of operational perception. It is beyond the scope of this chapter to explore the comparative advantages and weaknesses of each perspective. But if I were to do so, the preceding limitation in Merleau-Ponty would be balanced against his reflective appreciation of the numerous sensory "interinvolvements" that make perception possible. The starting point to engage Bergson on these issues is Merleau-Ponty (1962).

10 This book prompted me to take another look at Merleau-Ponty in relation to Foucault and Deleuze. Some will protest her assertion, saying that priority must be given either to the subject *or* to the object. But they then have to come to terms with the multiple interinvolvements elucidated by Merleau-Ponty and his judgment

that you cannot unsort entirely—once these mixings and remixings have occurred—exactly what contribution is made by one side or the other. Even the painter, alert to his powers of perception, is not "able to say (since the distinction has no meaning) what comes from him and what comes from things, what the new work adds to the old ones, or what it has taken from the others" (Merleau-Ponty 1964, 58–59).

11 That text also deepens our experience of the "flesh" in ways that extend all the points made about the sensorium discussed earlier. But we cannot pursue that pregnant topic here.

12 For a review of the neuroscience literature on bodily and cultural elements in the formation of sight, see Adam Zeman (2002, chapters 5 and 6).

13 In April 2005, the *Johns Hopkins Gazette* released the following bulletin: "Continuing its efforts to enhance the security of students, faculty and staff, the university has installed . . . a state of the art closed-circuit TV system. . . . The system can be programmed to look for as many as 16 behavior patterns and to assign them a priority score for operator follow-up. . . . The cameras are helping us to make the transition to a more fully integrated 'virtual policing' system."

14 Heath is not speaking of subliminal inserts here; he is talking about advertisements that distract attention from themselves and encourage the viewer to be distracted too as he inserts connections between affect, words, and images.

15 It is pertinent to emphasize that the "attachment to this world" spoken of here is not to existing injustices, class suffering, dogmatism, repression of diversity, and the like but to the human existential condition itself as it finds expression in a world in which some zones of life proceed at a rapid tempo. The wager is that the enhancement of attachment to this world increases the energy and will to oppose the dangers and injustices built into it.

16 I review strategies, both individual and collective, to rework dispositions to perception and sensibility in chapters 4, 5, and 6 of *Neuropolitics: Thinking, Culture, Speed* (Connolly 2001) and in "Experience and Experiment" (Connolly 2006).

References

Connolly, William E. 2001. *Neuropolitics: Thinking, Culture, Speed.* Minneapolis: University of Minnesota Press.

———. 2006. "Experience and Experiment." *Daedalus* 135, no. 3 (2006): 67–75.

Coole, Diana. 2000. *Negativity and Politics.* London: Routledge.

Damasio, Antonio. 2003. *Looking for Spinoza: Joy, Sorrow, and the Feeling Brain.* New York: Harcourt.

Deleuze, Gilles. 1989. *Cinema II: The Time Image.* Trans. Hugh Tomlinson. Minneapolis: University of Minnesota Press.

Deleuze, Gilles, and Félix Guattari. 1987. *A Thousand Plateaus.* Trans. Brian Massumi. Minneapolis: University of Minnesota Press.

Foucault, Michel. 1977. *Discipline and Punish.* Trans. Alan Sheridan. New York: Pantheon Books.

Goodwin, Brian. 1994. *How the Leopard Changed Its Spots.* Princeton, N.J.: Princeton University Press.

Gould, Stephen J. 2002. *The Structure of Evolutionary Theory.* Cambridge, Mass.: Harvard University Press.

Hansen, Mark B. N. 2004. *New Philosophy for New Media.* Cambridge, Mass.: MIT Press.

Heath, Robert. 2005. *The Hidden Power of Advertising.* Henley-on-Thames, U.K.: Admap.

Kelly, Sean D. 2005. "Seeing Things in Merleau-Ponty." In *The Cambridge Companion to Merleau-Ponty*, ed. Mark Hansen, 74–110. Cambridge: Cambridge University Press.

Marks, Laura U. 2003. "The Memory of Touch." In *The Skin of the Film*, 122–93. Durham, N.C.: Duke University Press.

Massumi, Brian. 2002. *Parables for the Virtual.* Durham, N.C.: Duke University Press.

Merleau-Ponty, Maurice. 1962. *Phenomenology of Perception.* London: Routledge and Kegan Paul.

———. 1964. "Indirect Language and the Voices of Silence." In *Signs*, trans. Richard McCleary, 39–83. Evanston, Ill.: Northwestern University Press.

———. 1968. *The Visible and Invisible.* Trans. Alfonso Lingis. Evanston, Ill.: Northwestern University Press.

———. 2003. *Nature: Course Notes from the College de France.* Trans. Robert Vallier. Evanston, Ill.: Northwestern University Press.

New Scientist. 2005. "Why You Have at Least 21 Senses." January 29.

Prigogine, Ilya. 2003. *Is Future Given?* Hackensack, N.J.: World Scientific.

Ramachandran, Vilayanur S. 1998. *Phantoms in the Brain.* New York: William Morrow.

Shapiro, Ian, Rogers M. Smith, and Tarek E. Masoud, eds. 2004. *Problems and Methods in the Study of Politics.* Cambridge: Cambridge University Press.

Sole, Ricard, and Brian Goodwin. 2000. *Signs of Life: How Complexity Invades Biology.* New York: Basic Books.

Zeman, Adam. 2002. *Consciousness: A User's Guide.* New Haven, Conn.: Yale University Press.

PART II

Technological Politics: Affective Objects and Events

4 Materialist Politics: Metallurgy

ANDREW BARRY

HOW MIGHT ONE CONCEIVE of the relation between materials and politics? As is common enough in science and technology studies, this chapter centers on a case study: the field of metallurgy and the materiality of metals and other inorganic matter. The danger of using a case study, of course, is that the case simply becomes an illustration of an idea or principle that has been formulated somewhere else—that which we already *know*, and that which we simply want to make *clear*—whereas what we would like from a case study is that it is something *more* than an example, that it tells us something that we do not know or creates an effect that is somehow unanticipated. The case should be placed in a setting where it can resist our explanations of it in some ways. There should be some irreducibility to the case. In other words, the case must make a difference.

There are, nonetheless, good reasons to use metals and metallurgy to think about the relations between materiality and politics. One reason is simply that there is something of a neglect of the politics of metals today, whether in terms of their extraction, manufacture or use, or repair. If the malleability of metals was once viewed as an index of the transformative capacities of capitalism, today metals seem to have disappeared from view. We live, according to many theorists, in a world marked by flows of knowledge and information, but materials are no longer of much interest. Where once they lay at the heart of social theory, metals appear to have been relegated to the backstage. In what follows, however, I am not concerned with the social shaping of metals; rather, I put forward a different thesis to the classical one, namely, that part of the political interest of metallurgy derives from its concern, as a form of field science, with the specificity of the case and the micropolitics of materials. My method is, to use Marilyn Strathern's (1995) term, *holographic*. Through a study of the politics of a field science that is concerned with the specificity of the case, I seek to illuminate why a concern with the specificity of the case is important for those concerned with the study of politics.

The chapter develops two arguments. One is that metals are not the hard, inert objects that they are often thought to be. Metals form part of dynamic, informed assemblages in which the expertise of metallurgists and other material and social scientists have come to play a critical part. They have become "informationally enriched," and part of the driving force for this informational enrichment comes from growing efforts to regulate the properties of the materials and the actions of those who develop and use them. The informational enrichment of materials, in short, has become a political matter. The second argument is that part of the political importance of metals and other inorganic materials derives from a sense that they have an objectivity and an immalleability that cannot be explained away as an expression of political ideology or economic interest. Defects in metals or accidents that derive from or lead to the failure of metallic and other material structures cannot easily be denied and cannot simply be viewed as a projection of the imagination of those who point to their occurrence. It is commonplace to stress the micropolitical importance of forms of creative, artistic, and inventive activity. Yet there is also a way in which natural scientific expertise, and its public performance, may also disrupt earlier certainties, fostering the emergence of new objects and sites of contestation. Through their expertise in the failure of metals and material structures, material scientists and metallurgists may themselves play an unexpectedly political part, turning the apparently mundane properties of specific materials and material structures, such as their fragility or toxicity, into issues of wider significance. The second part of this chapter focuses on an example in which a defect in a material structure, an oil pipeline, is understood to be an index of a much wider set of defects in corporate capitalism and its regulation. In this case a critique of the relations between corporate capital and government relied on the demonstrability of facts about the properties of specific materials. If political action often involves the staging of a particular issue as a matter of collective importance, then nonhuman materials rather than human subjects were, in this instance, placed firmly center stage (cf. Rancière 2004a).[1]

Metals

To begin, it would be a mistake to think that metallurgy is simply a branch of physics. Indeed, from the point of view of the metallurgist, the properties of metals cannot simply be deduced from fundamental physical principles.[2] Alloys cannot be understood as combinations of pure substances, and the behavior of metals in the conditions encountered in power stations or aircraft is quite different from any laboratory setting or simulation. Moreover, it would be a serious mistake to think that physics can simply be applied to the study of metals: or only if we take the word application to imply the need for a process, the path, the deviation, of translation (Callon and Latour 1981). One of the preoccupations of the metallurgist (and I use the term very broadly in this chapter to include all those concerned with the technical existence of metals and their relations to other substances[3]) is to be concerned with the specificity of the case rather than account for the case in terms of general principles. General principles are important, of course, but only so far as they are not applied in any generalized way and are acknowledged to be inadequate to the task at hand. The metallurgist expects that materials will be opaque, that the case will make a difference. In this way, the metallurgist is a good materialist, aware that materials will always, in some way, be resistant to external forces and will generate their own effects (Stengers 1997). Although not all may agree with this proposition, the socialist historian of science and crystallographer J. D. Bernal, writing in the early 1950s, reckoned that following the development of X-ray crystallography, it would be possible for metallurgists and other scientists to begin to take "rational control" over the internal structure of metals:

> The structural studies [following the development of X-ray crystallography] . . . explained the primary, economically valuable properties of metals—their plasticity and hardening, the means by which metals can be forged, rolled and drawn—and made possible the beginning of a rational control of these processes. (Bernal 1969, 796)

Though X-ray crystallography played a critical role in the development of molecular biology and solid state physics in the immediate postwar period, Bernal was overenthusiastic about the possibility of turning metals into what we might call, following Foucault, docile objects. After all, X-ray crystallography is a technique that can only be used to determine internal structural features of carefully prepared specimens in a well-equipped laboratory. It cannot be applied directly to the study of metals in use, or in the field, where it is likely that they will be subject to variations of stress or temperature and the effects of chemical action.

Insofar as metallurgy addresses the question of the relation between the transformation of metals and features of their external environment, it addresses a central problem for science and technology studies (STS). For STS was, of course, for a long time puzzled about the relation between external (economic and social) forces and the shape of technologies. In this way, STS rediscovered a classical problem (D. MacKenzie 1996). In a remarkable passage, Marx formulated the relation between the historical development of capitalism, the division of labor in manufacture, and the structure of metals, precisely in terms of their *shape:* "manufacture is characterised by the differentiation of the instruments of labour—a differentiation whereby tools of a given sort acquire fixed shapes, adapted to each particular application—and by the specialisation of these instruments, which allows full play to each special tool only in the hands of a specific kind of worker" (Marx 1973, 460).

Contemporary metallurgy does not confine itself to external form and shape, however; rather, one of the preoccupations of the metallurgist is with the question of how external forces and events become translated or absorbed at the level of molecular structure, and conversely, how molecular structure is mediated in transformations of external form. As Roux and Magnin argue, metallurgy is not so much the science of the microscopic *or* the macroscopic but a mesoscopic field that mediates between scales and spaces and between different forms and techniques of analysis (Roux and Magnin 2004, 11). The metallurgist is an expert who is capable of bringing different spaces and objects of analysis simultaneously into view, moving between observations of external and internal structure; between quantum physics, thermodynamics, corrosion chemistry, crystallography, and management strategy; between idealized atomic models and phase diagrams and materials in use; and between the human and nonhuman elements of assemblages (cf. A. MacKenzie 2002, 16).

From the point of view of contemporary metallurgy, metals are sites of transformation. Internally, they contain features, such as grain boundaries, regular lattice structures, impurities, dislocations, and catalytic sites, that provide the basis for both stability and rigidity *and* movement, elasticity and flow, and changes in intensive and extensive properties. They are spaces within which minute changes occur routinely, and catastrophic failures may represent the crystallization of a series of infinitesimal movements rather than the immediate impact of an external force (cf. Tarde 2001). It is common enough in social theory to draw an opposition between the static, the bounded or the rigid, and the fluid or the mobile. Indeed, for some, speaking of boundaries and rigidities at all is simply thought to be passé. But it would be wrong to oppose the solidity of metals with the fluidity of fluids or boundedness with flow; rather, it is a question of recognizing that solidity may itself be the product of a certain form of fluidity. After all, metals are extraordinarily fluid—full of local sources of transformation and instability—actually more fluid than fluids. Indeed, Deleuze and Guattari took the insights of the metallurgists to be an argument for vitalism: "what metal and metallurgy bring to light is a life proper to matter, a vital state of matter as such, a material vitalism that doubtless exists everywhere but is ordinarily hidden or covered, rendered unrecognisable" (Deleuze and Guattari 1987, 411). The metallurgist is not just concerned with the shape or mold within which metals are formed, or with their malleability, but with what can be termed the continuous modulation or variation of metals (Deleuze and Guattari 1987; see also Deleuze 1979).

So metals flow, and they share certain properties with living materials; it is just that they often flow more slowly, and from the point of view of the metallurgist, more profoundly and irreversibly than fluids. They can contain historical records of their past in a way that most fluids cannot. They have surfaces, but their surfaces are sites of transformation, such as corrosion and friction, as well as functioning as boundaries (Bowden and Tabor 2001). Metals' capacity to continue to exist over years and decades depends on fatigue and creep: the minute internal transformation of metals under fluctuating conditions of stress and temperature. So metals are quite unlike glass (which may shatter under the impact of an external force) or many fluids (which may simply move to another place, adapting to the shape of the container in which they are placed). Metals have the capacity to render external energies into novel internal forms, "modifying [themselves] through the invention of new internal structures" (Simondon 1992, 305). Metals are solid and hard and (for a period) can endure without ever remaining the same.[4] Their stability as material forms is intimately associated with both their internal transformation and their fragility (Roux and Magnin 2004).

But if metals have something of a metastable existence, passing slowly between states, they also come to exist in other forms generated through the work of metallurgists and the demands of regulators. In Bensaude-Vincent and Stengers's account of the *History of Chemistry*, instead of merely imposing a shape on matter, chemists proffer a "different notion of matter":

> Whether functional or structural, new materials are no longer intended to replace traditional materials. They are made to solve specific problems, and for this reason they embody a different notion of matter. Instead of imposing a shape on the mass of material, one develops an "informed material" in the sense that the material structure becomes richer and richer in information. Accomplishing this requires detailed comprehension of the microscopic structure of materials, because it is in playing with these molecular, atomic and even subatomic structures that one can invent materials adapted to industrial demands and control the factors needed for their reproduction, whether they are new or traditional. (Bensaude-Vincent and Stengers 1996, 206)

The same observation applies to metallurgy. The product of the contemporary metallurgist's labor is not necessarily a new metal; it is likely to be the "informational enrichment" of materials, multiplying their forms of existence. Through the work of metallurgists, metals acquire multiple lives: in simulations, micrographs, as X-ray crystallography, and samples taken from materials in use. In each of these settings, metals exist in different forms (more or less prepared, more or less purified, more or less isolated from other chemicals), which depend on particular informational–material practices of experiment and field research (cf. Mol 2002, 6; Barry 2005). Metals not only have a lively existence, their forms of existence increasingly depend on the informational–material assemblages through which they circulate. Moreover, although metallurgy might not provide the basis for the level of control of the properties of metals envisaged by Bernal, it nonetheless plays a critical role in their management and government. Consider, for example, the importance of the tests and measurements that are routinely carried out on systems such as power stations, aircraft, and oil platforms to ensure their integrity and safety. Such measurements are governmental acts: they are intended to manage the potentially unruly conduct of sociomaterial assemblages, aligning them with broader economic and governmental objectives. Just as the regulation of drugs demands the multiplication of their forms of existence as informed materials (through in vivo and in vitro investigations and through clinical and preclinical trials), metals are also subject to a series of commercial and regulatory tests, the results of which may or may not be made public (cf. Barry 2005; McGoey 2007).

Metallurgy might be described as something of a social and political science, if we understand the notion of the social in the sense given to it by the sociologist, Gabriel Tarde. For Tarde it was possible to refer to atomic or molecular societies as well as human societies, and he argued that the same concepts could refer to the societies described by the physical and life sciences as much as those analyzed by sociologists (Tarde 1999). The metallurgist might follow Tarde in acknowledging that there is no discontinuity between the realm of the social and the natural, the human and the nonhuman, or between the informational and the material, the living and the nonliving (Whatmore 2006; Barry and Thrift 2007; Thrift 2008). Whereas Bernal imagined that metallurgy would make it possible to establish something like a socialist administration of metals, resulting in a direct alignment between the internal structure of metals and economic need through the use of techniques such X-ray crystallography, contemporary metallurgists pursue a more flexible approach. For metallurgy assumes that there could be no correspondence between material and social and economic structure; rather, the metallurgist multiplies the forms in which metals exist, while recognizing that complete knowledge and control is impossible. Metallurgy is an interdisciplinary discipline, concerned with the study of systems, platforms, or processes, assemblages of which metals and other materials are only a part.[5]

Metals and metallurgy provide then a particularly good case study for thinking about the properties of materials. They clearly illustrate the principle of irreducibility: the behavior of metals resists any reduction of their properties, whether to their external (social) environment or to the fundamentals of physics. Metallurgists are mediators between the form of economic calculation, government regulation, and the analysis of material properties (cf. Osborne 2004). Moreover, metallurgy, like agricultural research, zoology, anthropology, and geography, is reliant on field research, a form of artisanal and itinerant practice (Deleuze and Guattari 1987, 411), and not just laboratory experimentation (Schaffer 2003; Livingstone 2003). As a field science, metallurgy should be, in principle, attuned to the specificity of the case. It is attentive to the general problem of how to address the study of the particular. Whereas many physicists, for example, may be preoccupied by the problem of how to represent the particular in terms of the general, metallurgists are often confronted by a rather different question, namely, how is it possible to understand and manage the properties of objects that exhibit general problems (such as fracture, conduction, phase transition, creep, or corrosion), but in specific ways and in very different settings and locations?

Politics

For Bernal the development of X-ray crystallography promised the possibility of a direct alignment between molecular and economic structures: the internal structure of metals would come to reflect economic needs. But how might we envisage the relation between failures of metals, or better, of specific sociomaterial assemblages, and more systematic failures in the economic and political order? This question is not hypothetical. In the recent past, the cases of Chernobyl and Three Mile Island, BSE and CJD, Brent Spar, GM crops and Bhopal, have come to be seen by many not just as accidents but as indices of more systemic failures. Failures in sociomaterial assemblages have been viewed as signs of the existence of a wider series of problems concerning the relations between science and politics and between governments, corporations, and citizens (e.g., Beck 1992; Berkhout et al. 2003). Their occurrence has demanded, according to many, a new politics of science and a new politics of the environment, science, and risk.[6] Particular accidents have come to be understood as markers of general problems and as specific cases that demand general solutions and wider forms of political action. In short, particular failures have been constituted as political events.[7]

In these circumstances, how might one envisage the relation between metallurgy, metals, and politics? If metallurgy is an example of a field science, an itinerant practice itself entangled with the study of the particular in its environment, then how do the insights of metallurgists come to have more general significance? What is the relation between the claims made by metallurgists about the specificity of the particular and the claims by those with an interest in politics concerning the relations between particular occasions and collective issues and concerns? In what follows I focus on an example of an occasion in which metallurgy plays a remarkable and unexpected role. The case raises two questions. The first concerns the relation between the properties and behavior of metals and the organization of political and economic life. How can particular material processes, including accidents, come to be constituted as events of general significance to others? And in what circumstances are they not? Second, why might political controversy focus on the behavior of metals and other nonhuman substances rather than the behavior of humans? And how can the work of metallurgists be made to have such potent political agency?[8]

The case in this chapter is an enquiry by Parliament's House of Commons Select Committee on Trade and Industry into the activities of the U.K. government's Export Credit Guarantee Department (ECGD) in 2005. In particular, it focused on the operation of the department's Business Principles, which were expected to govern the relation between the department and the companies to which it provided financial assistance. Yet although the enquiry had a very specific focus and examined the activities of particular and arguably minor government agencies, these activities raised, according to critics, wider questions. Did the government exercise control over the behavior of corporations in other countries, or does the government primarily act to facilitate corporations' activities? What is the character of relations between the government and multinational corporations? Or, even more broadly, are the Business Principles of the British government simply particular features of the operation of neoliberalism or the "neoliberal state" (Harvey 2005)?

Though the remit of the select committee was to address the implementation and effectiveness of the Business Principles by the ECGD, it nonetheless came to focus on a particular example of the implementation of these principles. This was the financial support given by the ECGD, in conjunction with the International Finance Corporation (IFC) and the European Bank for Reconstruction of Development (EBRD), for the construction of the Baku-Tbilisi-Ceyhan (BTC) oil pipeline, one of the largest single construction projects in the world in the early 2000s.[9] The development of the pipeline had been promoted in the 1990s by both the Turkish government and the U.S. Clinton administration as a way of bringing oil from the Caspian Sea along a route, through Azerbaijan, Georgia, and Turkey, that avoided both Iran and Russia. At the same time, this explicitly geopolitical investment would serve to bring Azerbaijan, and possibly other Turkic-speaking republics of the former Soviet Union, into the U.S. or Turkish sphere of influence. The involvement of the IFC, EBRD, and ECGD in the project was intended to reduce the financial risk to investors but also helped to ensure that the U.S. and U.K. governments, in particular, would have a direct interest in the completion of the project.[10] Oil companies were willing to be submitted to the greater scrutiny that the receipt of public finance would entail, in part because it would ensure that Western governments would have this interest.

However, even within this restricted focus on the financial support of the ECGD for the BTC pipeline, the select committee channeled its critical scrutiny still further. Prompted by the work of a coalition of nongovernmental organizations (NGOs), critical of the work of the ECGD, the committee devoted considerable time to the failure of a particular coating material used on joints between sections of the pipeline.[11] More precisely still, it was concerned with the very specific issue of what the ECGD knew about the procurement and use of this coating material in late 2003, during the period when the department was considering whether to support the construction of the pipeline. Indeed, the case of the coating material was the *only* issue related to the BTC pipeline discussed in the House of Commons, with the exception of a brief discussion of the case of the Kurdish nationalist activist Ferhat Kaya, who was allegedly tortured in the police station in the Turkish town of Ardahan, near the Georgian border, on account of his criticism of the BTC project (Bakuceyhan Campaign 2004; House of Commons 2005b, 52).

The centrality of this particular coating material to the concerns of British members of Parliament is moreover surprising when viewed in relation to debates elsewhere. For a time, during 2003–5, the pipeline acquired a remarkable political geography. In Washington, D.C., in particular, BTC came to have a very different significance. The offices of the ombudsman of the International Finance Corporation on Pennsylvania Avenue, for example, investigated a series of specific alleged violations of bank guidelines by the BTC company in Georgia, following representations made by the Georgian environmental NGO Green Alternative (2004). These concerned, for example, the alleged failure of the BTC company to ensure adequate compensation to villagers whose houses had been damaged by subcontractors. Elsewhere in Washington, the State Department was forced to intervene following the decision of the Georgian government of Mikheil Saakashvili to temporarily halt construction of the pipeline in July 2004, and the issue was discussed in meetings between Saakashvili and Colin Powell and Donald Rumsfeld.[12] Moreover, the failure of the coating material described by the metallurgist had not led to any oil leak or effect on the environment. There was no specific accident to which anyone could point, although NGO critics described it as an "environmental time bomb" and linked it, by association, to a series of events involving BP, including connections between the oil company and Colombian paramilitaries (House of Commons 2005b, 105). Nor did the problem have any discernable impact on the complex geopolitical situation within which the pipeline was embedded. The failure of the coating material was not considered of particular importance by villagers living near the pipeline route, who were incensed by their failure to receive compensation that they had expected to receive because of the presence of oil industry construction work near their homes.[13] In a meeting with Georgian workers and residents in the city of Rustavi nearby to the pipeline route, I was told that up to fifty kilometers of pipeline had to be relayed.[14] But this was of little concern to the workers, who were angry about low pay, long working hours, and poor food, and who had been engaged in unofficial strike action in the same period. However, the issue of working conditions and wages was not considered by the select committee, even though it might reasonably have done so. After all, the working and wages of the Georgian pipeline workers were governed by the Host Government Agreement between the BTC company and the Georgian government, which allowed for the pipeline *not* to be governed by some of the conditions of Georgian labor law. This agreement appeared to violate the terms of the Organisation for Economic Co-operation and Development (OECD) Guidelines on Multinational Enterprises, which stipulate that multinationals should not seek or accept exemptions from the provisions of local law (Organisation for Economic Co-operation and Development 2000, 19). It was, potentially, a good example of neoliberal government in practice. The question of this particular exemption from the guidelines, however, was not considered an issue, in public at least, either in London or Washington, D.C.[15]

Why, then, should this committee, prompted by NGOs campaigning against the pipeline, take such particular interest in these cracks and the specific issue of pipeline coating material rather than the working conditions of Georgian pipeline workers or the partial exemption of BTC from the terms of Georgian labor law, for example? Why was the politics of an "informed material" considered more significant than the politics of class (Gibson-Graham 2006)? If politics, as Rancière suggests, involves making objects and problems visible, why were the failures of material objects rather than the working conditions of laborers rendered visible to the committee (Rancière 2004b, 226)? Why should defects in materials—rather than defects in labor relations, pay, and working conditions—stand in for wider problems between business and the U.K. government or, more generally, between state and capital? Why, in this case, did the properties of materials come to have such political significance (Barry 2001, 215)?

An answer to these questions is complex. For if metallurgy is a form of field research that needs to address the specificity of the case, the same is true of field research concerned with the study of politics. An analysis of this event would involve consideration, for example, of the critical historical role of the Green Movement in both Soviet and post-Soviet Georgian politics.[16] It would involve examination of the particular timing of the failure of the coating material that occurred just before the decision of the ECGD to support the development of the pipeline. It would involve an analysis of how the question of the reputation of oil corporations has become a focus for both management and political action, particularly following public criticism of Shell concerning the disposal of the Brent Spar oil platform in 1995 (Power 2007, 128–29). Crucially, it would involve an assessment of the preoccupation with formal procedures of accountability and transparency in political and economic life (Power 1997; West and Sanders 2003; Best 2005). In these circumstances, the production of information about materials, as much as the production of information about labor relations or human rights, can in principle become a public political matter.

The salience of the politics of materials, rather than the politics of labor in the House of Commons, also turns partly on the legitimacy of particular sources of evidence. After all, the failure of the coating material could not be denied, for everyone, including BP, accepted that it had happened. Long sections of pipeline had to be repaired as a consequence of the failure of the coating material. Once acknowledged, this could not be simply explained away by any suggestion that the failure of the coating material was conjured up by the opponents of the oil company or the government for political or financial gain. Unlike the material demands by pipeline workers that they should work shorter hours and be paid at higher rates, claims concerning the existence of cracks could not so easily be accused of being self-interested or, indeed, even "politically motivated." In comparison to the protests of the pipeline workers, the materiality of cracks in the pipeline coating material was less clearly entangled in the complexities of Georgian politics in the aftermath of the Rose Revolution.[17] And unlike the demands of Georgian workers, which were mediated by local lawyers and trade union representatives who did not speak English, the existence of cracks was mediated directly in London through the work of well-funded international NGOs.[18]

In what follows, however, I focus more narrowly on the question of the presentation of evidence in the House of Commons. After all, the potential significance of evidence depends on the setting in which the evidence is presented and the audience to whom it is presented (Shapin and Schaffer 1985). In representing evidence of the failure of pipeline coating material in the House of Commons, radical NGOs sought to effect a radical translation in its significance. Evidence of the existence of material failure mattered in the House of Commons not primarily because it involved information about materials, and their local conditions of existence in use, but because of NGOs' sense of the materiality of this information in relation to the behavior of the government and the multinational. Critics expected that evidence presented in the House of Commons would have a quasi-legal effect, demonstrating the guilt of the multinational and its supporters in government in a public forum. In this setting, the particular was of little interest in terms of its particularity, but in terms of how far it could be seen as a manifestation of the wider forms of complicity between corporate business and government. It provided the basis for an empirical critique of the capitalist state, one might say, pointing to the existence of a network of relations between officials and businesspeople that otherwise would be unacknowledged.[19] But how was it possible to translate knowledge of the behavior of informed materials in a specific locality, of no obvious significance to a group of parliamentarians, into information that was of material importance to the recommendations of a select committee? How could one translate a (technical) fact about the failure of materials in the field into a (quasi-legal) fact that would matter to the deliberations of a select committee and demonstrate the guilt of the government and the multinational (Latour 2004)?[20] Critical to the NGOs' case before the select committee was the testimony of a metallurgist concerning the period prior to the start of pipeline construction in 2003. This testimony was expected to acquire political agency once presented in the House of Commons.

In November 2003, shortly after the Rose Revolution in Georgia that led to the end of the government of Eduard Shevardnadze, cracks in the material that covered the connections between separate sections of pipe emerged during the construction. The BTC company claimed that the cause of the fault was that the field joint coating covering the connections had been misapplied as the temperature dropped in November, but that following further investigations and tests, the problem had been rectified. Despite the previous existence of cracks in the coating material, the pipeline could be buried safely. The metallurgist, himself a consultant who had offered his services to BP, the major oil company involved in the BTC project, was incensed that the company had previously failed to think through the relations between their actions in selecting this particular coating material for the oil pipeline and the behavior of the pipeline in the field. The metallurgist explained to the parliamentarians:

> METALLURGIST, *reading from a report commissioned by BP concerning the field joint coating material:* ["]The coating may or may not be damaged in cold weather, but it will certainly not suffer the same damage from soil stressing as the alternatives available.["]
>
> I cannot believe the crassness of these statements. They are saying that they did not know if the joint coating would be damaged, or not, during backfilling—absolutely astounding! But then of course they could always find out "on the job," another example of the "guinea pig" engineering culture.
>
> Then they say definitively that it will not suffer soil stressing as badly as alternatives—when they did not test any of these alternatives. This is the judgement of the crystal ball! It is certainly not an engineering judgement. The fact is that had the joint been coated with a mimic three layer system employing injection moulded PE top

> coat, the field joint would actually have had a superior soil stressing resistance. . . .
>
> Little or no reference is made in the WP [i.e., the BP] report with regard to in-ground performance of the epoxy yet this is fundamental to the coating's ability to protect the pipe in the long term.
>
> Oil and gas pipelines are not passive, inert items, they are *live, dynamic structures* that move due to ground movement and most importantly, pressure changes within the pipe. . . . The coating has to accommodate such movement. The operating temperature will fluctuate with pressure changes and should the pipeline be shut down for any time, the pipe temperature will drop down to the in-ground ambient—estimated by BP to be −5° to +50° C. . . . How will this affect the performance of the coating particularly at the PE/epoxy interface . . . ? This question has been discussed throughout the whole pipeline industry and I am yet to hear any individual say—"it will be OK, the system is fully proven." (House of Commons 2005b, 61)

The metallurgist argued, furthermore, that the modified epoxy coating had been inadequately tested, that the specification for the coating was inadequate, that documentation was unsatisfactory, and that tried and tested alternatives were not properly considered. In short, using SPC2888 involved a considerable and unnecessary risk: "if you have something that does the job and these other systems have been extensively applied and [have] a working history why change and in particular to use this very important pipeline as a proving ground for an experiment with a new coating system" (House of Commons 2005b, 85). Earlier he had warned BP, "Have you considered the insurance implications of this?" (House of Commons 2005b, 81). For the NGOs and a journalist, the defects in SPC2888 embodied defects in BP itself and its relations with ECGD and the lender's group consultants on whom ECGD relied in their exercise of due diligence. These consultants, according to the journalist, were, in effect, told by the lenders to rely on the integrity of BP in providing them with accurate information. This was a scandal: due diligence assumed that the company could be trusted even when there were those who were able to provide evidence to show why it should not be. The failure to investigate defects in coating material reflected wider defects in the activities of multinationals, banks, and government and their all-too-intimate relations:

> This statement [that the lender's group did not want the problem examined further] provides an extraordinary insight into the approach taken by the Lenders group [including the ECGD] after its much vaunted due diligence procedures were exposed by the *Sunday Times*. They went on to limit the investigation of the field joint coating issue to a simple desktop study. (House of Commons 2005b, 105)

But if the metallurgist's willingness to speak openly about his concerns with BP provided the opportunity for radical NGOs to demonstrate the complicity of multinationals and government, his extraordinary statement to the House of Commons points to a very different kind of politics, and a different form of expertise, to that of the NGO critics. If politics partly revolves around the question of how the particular is figured as an instance of interest to a collective, then the metallurgist's political concerns, and his understanding of the relation between the particular and the general, are quite distinct. For although the metallurgist spoke of cracks in coating materials, he viewed these as an index of a "guinea pig engineering culture" that failed to attend to the liveliness of materials rather than as a sign of political complicity. Nor would he imply a link, for example, between the oil company's poor quality-control procedures and the association of its activities with human rights abuse by the Turkish police, evidenced by the case of Ferhat Kaya. For NGO critics of the multinational, the torture of Ferhat Kaya and the failure of the coating material, along with a whole series of other specific events and incidents, were considered signs of the state of relations between government and the oil business.

Although the metallurgist gave evidence, he also gave his evidence with passion and anger. In so doing, he gave up the pretence that his evidence was, as the evidence of a scientist might be expected to be, dispassionate (Bennington 1994, 135). His anger derived partly about how badly particular elements—this steel, this soil, this coating material, the skills of these subcontractors, the winter climate of

Georgia, and so on—had been assembled together. And he detailed the reasons why this occurred, with the specificity of this case that had so many surprising wider consequences. Metallurgy here stands as an example of an itinerant and artisanal practice that, potentially at least, addresses the impossibility of fully governing the behavior of materials, taking proper notice of their differential resistance. The metallurgist was not surprised by the failure of materials because materials are not the dead, inert substances they are sometimes imagined to be. Nor was he disinterested and unaffected. The intensity of the metallurgist's anger, expressed in Parliament, stemmed from his belief that the oil company had put such a badly formed assemblage together. It had tried something out without having properly checked to see if it was going to work.[21] The metallurgist entered into the unfamiliar terrain of public politics not because it was in his interests to do so (it almost certainly was not, and he claimed to have become ill as a result of his intervention), nor because of his anticorporate politics (there is no reason to suppose that he had these). His preoccupation was with the irreducibility of the properties of metals and a defense of the autonomy of his modest expertise of the behavior of an informed material. His was a more-than-human politics (Whatmore 2006).

The significance of the metallurgist's testimony was judged in a public setting: the select committee (cf. Lynch 1998; Schaffer 2005). Within Parliament, select committees have a particular significance. As in the U.S. Congress, a select committee is a group of politicians, selected from all parties, who interrogate the conduct of government and the development and implementation of legislation in public. A parliamentary committee is not a court of a law, for its recommendations do not carry the force of law. Nor is it a community of experts, for although a select committee may seek expert advice and is likely to have its own expert advisor, it does not claim any expertise itself. Yet, like a court of law, a select committee is expected to function as a space where matters of fact can be established and judgments can be made on the basis of the evidence presented before it (cf. Latour 2004). Moreover, on account of the authority of Parliament, it is able to request evidence and witnesses who may not be available otherwise and who, with exceptions, are required to give evidence in public. However, unlike the main chamber of the House of Commons, its final recommendations are expected, in general, to reflect the views of all of its members and not just the views of the governing party or the statistical majority of the members of Parliament (Waldron 1999, 127). In this way, a select committee is potentially in the position to claim that its views are based on consideration of evidence and, at the same time, to be able to articulate, in principle, a nonparty political agreement based on this consideration. Perhaps more than any other parliamentary institution, parliamentary committees claim to be able to act as "modest (political) witnesses": ladies and gentlemen who confront evidence with disinterest (Shapin and Schaffer 1985; Latour and Weibel 2005) and yet who also represent the public interest. In effect they are thought to perform a function, regarded as essential in the institution of British parliamentary democracy, that it is possible to reach an agreement, not through consensus, and despite underlying disagreement, given the existence of an appropriate institutional mechanism and the prevalence of a certain form of ethical conduct in political life. At the same time, they were concerned to judge not just the veracity of the metallurgist's statement but whether it was a matter of public concern. Should the failure in materials be an index of a wider failure in the relations between business and government? Should it even become an event that inaugurated a transformation in these relations?

Despite their exhaustive preoccupation with the circumstances surrounding the failure of SPC2888, the parliamentarians ultimately were unconvinced about its wider significance. After all, their concern was with the behavior of ECGD in relation to BP and its adherence to its Business Principles, not with the conduct of BP itself. The domain of the market economy (BP) was considered outside the domain of politics (Barry and Slater 2005). "It was not surprising," according to the select committee, "that quality assurance problems occur during major construction projects such as the BTC pipeline. What matters is that those problems are identified and addressed" (House of Commons 2005a, 12). For the parliamentarians, the ECGD and the government had done all they could reasonably do to ensure that the problem of the pipeline coating was addressed: the ECGD had taken "proportionate and consistent action" (House of Commons 2005a, 13). They had done enough to investigate the properties of SPC2888. As MPs they were not in a position to make a judgment about the behavior of materials, only about the behavior of government. And they based their judgment, in the manner of a court, not on the commissioning of a piece of independent field research on the situation in Georgia but on the basis of evidence presented before them (cf. Latour 2004, 101).

Nonetheless, there is no simple explanation for the parliamentarians' decision.[22] To account for the decision, one would need to consider the particular composition of the committee and its relations to government ministers, for example, and the level of trust of parliamentarians in BP in comparison to other U.K. companies. And one would need to examine the work of other metallurgists commissioned by both BP and the ECGD and the evidence they provided. The metallurgist's evidence was, after all, but one of a number of published and unpublished reports of the performance of the pipeline that circulated between Georgia and BP and government offices in Baku and London (cf. Bridge and Wood 2005). There is, moreover, the question of whether a scientist, who expressed his views with such anger, was trusted by those who listened to his testimony. But in my reading, part of the reason why the evidence of the metallurgist was not thought to be a matter of wider concern is the way in which his intervention was read too politically by both politicians and NGOs. In effect, he was viewed as an agent or an instrument of an explicitly political campaign against the government and the oil company. In this way, his concern with the specificity of materials, and the particular location and manner of their use, was understood too readily within a given political context. His micropolitics, which relied on his own understanding of the dynamic behavior of informed materials, was overinterpreted in macro or molar political terms (cf. Deleuze and Guattari 1987, 216; Barry and Thrift 2007, 514). In this situation, the failure of materials and the metallurgist's evidence concerning this failure could not be made to matter beyond the confines of Parliament.

Conclusions

Radical critics of capitalism have often developed their arguments either through an analysis of capitalism's systemic features and/or by making visible, through specific cases, the forms of human misery, inequality, and exploitation that are associated with capitalism's development. General analyses of capitalism's systemic features have framed particular accounts, and specific examples have been taken as indexes of systemic problems. In the case discussed here, the failure of material structures was taken by radical critics as a sign of wider defects in the relations between government and business. This was a critical strategy grounded in a form of legal empiricism.

Yet if the behavior of materials is sometimes taken to be an index of wider social relations, there is nothing naturally political about metals or other materials or how they are shaped. If one common feature of political life is that specific issues or problems are made (for a time and in particular settings) into matters of collective or "universal" significance (Žižek 2004, 70; Runciman 2006), and thereby become political, then there is no necessary reason why the behavior or properties of specific materials should be considered a political matter. To be sure, forms of critical analysis help them to become so, yet such critical analysis can also interpret the political significance of materials in reductive ways (Mitchell 2002, 52). It is not inevitable that the behavior of materials should be of interest to others or be the object of disagreement across a range of sites and settings within which political matters are addressed, whether in public or not. Materials acquire more-than-local political agency only occasionally, not in general.

The political importance of metals and metallurgy arises therefore in particular circumstances and sites. In this case, it depended on the coincidental timing of a stage in a decision-making process (whether to provide financial support for the construction of a pipeline) with a material event (the emergence of cracks in the pipeline coating material). It depended on the behavior of metals and liquid epoxy coating materials when applied in freezing conditions. It depended on the progressive formation of London as a center of expertise and political debate concerning the question of corporate social responsibility in recent years. It depended on the preoccupation with formal processes of accountability, transparency, and reputation in contemporary political and economic life, which made it possible for both an oil company and a government department to be accused of failing to be transparent, and for this to be considered potentially a matter of public political interest. And it depended on the existence of a parliamentary political assembly which, for a period, became interested to hear evidence of the complicity between government and business.

In these circumstances, the analysis of political events needs to attend to the timing and spacing of political life, the moment and setting of politics, and the specificity of its techniques, institutions, forms of evidence, and speech. But it should also address the ways in which the behavior of metals and other materials plays a critical part in politics. Metals are not the inert objects they are sometimes imagined to be, merely shaped by social and economic forces. They are elements of lively dynamic assemblages that may act in unanticipated ways, serving as the catalyst for political events. Metallurgists are well aware of the difficulty of applying the general principles of physics and chemistry to particular cases and of the need to recognize the unpredictability of material processes and the fragility of materials. Metallurgy is a form of artisanal and itinerant practice that needs to attend to the specificity of the case. These lessons are also relevant to those concerned with the study of politics.

Notes

My thanks to Bruce Braun, Georgina Born, Alberto Toscano, and Sarah J. Whatmore for their comments on an earlier draft of this chapter.

1 On the constitution of nonhuman entities as political issues, see Barry (2001) and Marres (2005).
2 In this respect, this chapter follows others that argue for the need to overturn the conventional hierarchy of the disciplines that places "fundamental" sciences (physics, molecular biology, and the neurosciences) at the top of the hierarchy and less fundamental disciplines, including chemistry, agronomy, metallurgy, physical geography, and social anthropology, further down (Schaffer 2003; Stengers and Bensaude-Vincent 2003; Barry 2005).
3 In the chapter I leave aside the question of how the relation between metallurgy and the broader field of materials science is conceived

by actors. Metallurgy, along with materials science more broadly, is in any case an interdisciplinary field that incorporates elements of chemistry, physics, crystallography, and indeed management theory (see note 3). On the broader question of the interdisciplinarity of disciplines, see Barry, Born, and Weszkalnys (2008).

4 Whitehead uses the example of the mountain to explain endurance as a process of transformation: "The mountain endures. But when after ages it is worn away, it has gone" (Whitehead 1985, 107).

5 "In industry, it is rarer to see a 'materials department,' rather technical departments will now tend to be identified by the product—or in the aerospace sector as the 'system' or 'platform.' An aeroengine is a system in this sense, and the technical team will involve materials scientists alongside aerodynamicists, structural engineers, electrical engineers, designers, etc. [In] university research, we are moving slowly to this systems approach, or 'interdisciplinary' research as it is more normally called in the academic sector. Many of the modern challenges in materials are not solely about 'new' materials, but rather materials integration into systems with specified overall function" (P. Grant, pers. comm., 2007).

6 David Runciman provides an elegant analysis of the constitution of 9/11 as an event of world historical importance (Runciman 2006).

7 The constitution of an occasion or an accident as an event depends, of course, on its mediation by others (cf. Barry 2002; Dewsbury 2007).

8 This question is posed by Timothy Mitchell (2002, 53): "[An analysis of human agency] means acknowledging something of the unresolvable tension, the inseparable mixture, the impossible multiplicity, out of which intention and agency must emerge. It means acknowledging that human agency, like capital, is a technical body, is something made."

9 ECGD provided up to $150 million cover for the project (House of Commons 2005a, 8).

10 This chapter draws on research for an Economic and Social Research Council–funded project on "Social and Human Rights Impact on the Governance of Technology" (2004–5). The research involved officials of the World Bank, the U.K. government, and the European Bank Reconstruction and Development, and four periods of fieldwork in Turkey, Azerbaijan, and Georgia in 2004.

On the geopolitics of oil development in the Caspian region in the immediate post-Soviet period, see Croissant and Aras (1999) and Ebel and Menon (2000).

11 Cornerhouse describes itself as a group that aims to support democratic and community movements for environmental and social justice through research and advocacy. Its approach is based on evidence: "we try to take a 'bottom-up' approach, filled with examples, to issues of global significance which are often handled in a more abstract way" (http://www.thecornerhouse.org.uk).

12 "Rumsfeld Intervention Rescues $3bn BP Pipeline," *Independent*, August 9, 2004.

13 On the question of the relation between expectation and affect, see Anderson (2006).

14 Field notes, April 2004.

15 International and national NGOs did, however, raise a series of other issues concerning the terms of the Host Government Agreement. However, the Georgian NGO that scrutinized the text of the agreement worked with a Georgian translation and appeared not to have noticed this particular exemption of the pipeline from the terms of Georgian law (Abashidze 2003). Although the construction of the BTC pipeline was subject to unprecedented levels of monitoring and thousands of pages of documentation were published about the environmental and social impacts of the pipeline, there is very limited public information about workers' wages and conditions. By contrast, the level of detail available to Marx (1973) through the reports of the Inspectors of Factories in the mid-nineteenth century is considerable.

16 The Georgian Green Movement was founded as early as 1988 (Wheatley 2005, 48). One of its first leaders, Zurab Zhvania, was prime minister (2004–5) in the Saakashvili government. In comparison to Georgia, political interest in environmental issues is undeveloped in neighboring countries, including Azerbaijan and Turkey.

17 E.g., it was rumored that the workers' protest was instigated by politicians opposed to Saakashvili. Whether this was true, cracks in materials could not so easily have been accused of being so politically motivated.

18 On the role of mediators, see Osborne (2004).

19 In this respect the strategy of the NGOs bares comparison with the empirical critique of the capitalist state provided by Ralph Miliband,

who pointed to the existence of specific networks of relations between government and business: "the world of administration and the world of large-scale enterprise are now increasingly linked in terms of an almost interchanging personnel" (Miliband 1973, 112). Miliband was famously criticized by Poulantzas for his narrowly empirical focus on human agents, which failed to account for the structural conditions of state action (Jessop 1990, 250).

20 As Bruno Latour notes, the word *fact* means something quite different in science and the law: "rather than confuse the two, we should sharpen the contrast: when it is said that the facts are there, or that they are stubborn, that phrase does not have the same meaning in science as it does in law, where, however stubborn the facts are, they will never have any real hold on the case as such, whose solidity depends on the rules of law that are applicable to the case" (Latour 2004, 89). While the operation of a select committee has some similarities to a court of law, it is a distinct form of political assembly, the characteristics of which have yet to be investigated.

21 The work of the metallurgist is an indicator of the complex geography of knowledge production in the oil industry, which relies on the production of a whole series of different forms of knowledge that may be more or less attuned to the existence of local specificities (Bridge and Wood 2005, 206).

22 On the idea of the "decision," see Law (2002, 143–62).

References

Abashidze, Besarion. 2003. *Baku-Tbilisi-Ceyhan: The Legal Analysis of the Host Government Agreement*. Tbilisi: Georgian Young Lawyers Association.

Anderson, Ben. 2006. "Becoming and Being Hopeful: Towards a Theory of Affect." *Environment and Planning D* 24, no. 5: 733–52.

Bakuceyhan Campaign. 2004. Statement of Ferhat Kaya Regarding Arrest and Ill-Treatment Following Work on BTC Pipeline. http://www.bakuceyhan.org.uk/ferhat_statement.htm.

Barry, Andrew. 2001. *Political Machines: Governing a Technological Society*. London: Athlone.

———. 2002. "The Anti-political Economy." *Economy and Society* 31, no. 2: 268–84.

———. 2005. "Pharmaceutical Matters: The Invention of Informed

Materials." *Theory, Culture, and Society* 22, no. 1: 51–69.
Barry, Andrew, and Don Slater, eds. 2005. *The Technological Economy*. London: Routledge.
Barry, Andrew, and Nigel Thrift. 2007. "Gabriel Tarde: Imitation, Invention, and Economy." *Economy and Society* 36, no. 4: 509–25.
Barry, Andrew, Georgina Born, and Gisa Weszkalnys. 2008. "Logics of Interdisciplinarity." *Economy and Society* 37, no. 1: 20–49.
Beck, Ulrich. 1992. *The Risk Society: Towards a New Modernity*. London: Sage.
Bennington, Geoffrey. 1994. *Legislations: The Politics of Deconstruction*. London: Verso.
Bensaude-Vincent, Bernadette, and Isabelle Stengers. 1996. *The History of Chemistry*. Cambridge, Mass.: Harvard University Press.
Berkhout, Franz, Melissa Leach, and Ian Scoones, eds. 2003. *Negotiating Environmental Change: New Perspectives from Social Science*. Cheltenham, U.K.: Edward Elgar.
Bernal, John D. 1969. *Science in History*. Vol. 3. *The Natural Sciences in Our Time*. Harmondsworth, U.K.: Penguin.
Best, Jacqueline. 2005. *The Limits of Transparency: Ambiguity and the History of International Finance*. Ithaca, N.Y.: Cornell University Press.
Bowden, Frank P., and David Tabor. 2001. *The Friction and Lubrication of Solids*. Oxford: Oxford University Press.
Bridge, Gavin, and Andrew Wood. 2005. "Geographies of Knowledge, Practices of Globalization: Learning from the Oil Exploration and Production Industry." *Area* 37, no. 20: 199–208.
Callon, Michel, and Bruno Latour. 1981. "Unscrewing the Big Leviathan: How Actors Macrostructure Reality and How Scientists Help Them Do So." In *Advances in Social Theory and Methodology*, ed. Karin Knorr-Cetina and Aaron Cicourel, 277–303. London: Routledge and Kegan Paul.
Croissant, Michael, and Bulent Aras, eds. 1999. *Oil and Geopolitics in the Caspian Sea Region*. London: Praeger.
Deleuze, Gilles. 1979. "Metal, Metallurgy, Music, Husserl, Simondon." http://www.webdeleuze.com.
Deleuze, Gilles, and Félix Guattari. 1987. *Thousand Plateaus: Capitalism and Schizophrenia*. Minneapolis: University of Minnesota Press.
Dewsbury, J. D. 2007. "Unthinking Subjects: Alain Badiou and the Event of Thought in Thinking Politics." *Transactions of the Institute of British Geographers* NS 32: 443–59.

Ebel, Robert, and Rajan Menon, eds. 2000. *Energy and Conflict in Central Asia and the Caucasus*. Lanham, Mass.: Rowman and Littlefield.

Gibson-Graham, J. K. 2006. *A Postcapitalist Politics*. Minneapolis: University of Minnesota Press.

Green Alternative. 2004. "Complaint to IFC Ombudsman." http://www.cao-ombudsman.org/cases/document-links/documents/BTC-FinalAssessmentReport_SevenComplaints_English_001.pdf.

Harvey, David. 2005. *A Brief History of Neo-Liberalism*. Oxford: Oxford University Press.

House of Commons. 2005a. *Implementation of ECGD's Business Principles: Ninth Report of the Session 2004–5, vol. 1, Report Together with Formal Minutes,* HC 374-I.

House of Commons. 2005b. *Implementation of ECGD's Business Principles: Ninth Report of the Session 2004–5, vol. 2, Oral and Written Evidence,* HC-374-II.

Jessop, Bob. 1990. *State Theory: Putting Capitalist States in Their Place*. Cambridge: Polity.

Latour, Bruno. 1999. *Pandora's Hope: Essays on the Reality of Science Studies*. Cambridge, Mass.: Harvard University Press.

———. 2004. "Scientific Facts and Legal Objectivity." In *Law, Anthropology and the Constitution of the Social,* ed. Alain Pottage and Martha Mundy, 73–114. Cambridge: Cambridge University Press.

Latour, Bruno, and Peter Weibel, eds. 2005. *Making Things Public: Atmospheres of Democracy*. Cambridge, Mass.: MIT Press.

Law, John. 2002. *Aircraft Stories: Decentering the Object in Technoscience*. Durham, N.C.: Duke University Press.

Livingstone, David. 2003. *Putting Science in Its Place: Geographies of Scientific Knowledge*. Chicago: University of Chicago Press.

Lynch, Michael. 1998. "The Discursive Production of Uncertainty: The OJ Simpson 'Dream Team' and the Sociology of Knowledge Machine." *Social Studies of Science* 28, nos. 5–6: 829–68.

MacKenzie, Adrian. 2002. *Transductions: Bodies and Machines at Speed*. London: Continuum.

MacKenzie, Donald. 1996. *Knowing Machines: Essays on Technical Change*. Cambridge, Mass.: MIT Press.

Marres, Noortje. 2005. "No Issue, No Publics: Democratic Deficits and the Displacement of Politics." PhD diss., University of Amsterdam.

Marx, Karl. 1973. *Capital*. Vol. 1. Harmondsworth, U.K.: Penguin.

McGoey, Linsey. 2007. "On the Will to Ignorance in Bureaucracy." *Economy and Society* 36, no. 2: 212–35.

Miliband, Ralph. 1973. *The State in Capitalist Society.* London: Quartet.

Mitchell, Timothy. 2002. *Rule of Experts: Egypt, Techno-Politics, Modernity.* Berkeley: University of California Press.

Mol, Annemarie. 2002. *The Body Multiple: Ontology in Medical Practice.* Durham, N.C.: Duke University Press.

Organisation for Economic Co-operation and Development. 2000. *The OECD Guidelines for Multinational Enterprises.* Paris: OECD. http://www.oecd.org/dataoecd/56/36/1922428.pdf.

Osborne, Thomas. 2004. "On Mediators: On Intellectuals and the Ideas Trade in the Knowledge Society." *Economy and Society* 33, no. 4: 430–47.

Power, Michael. 1997. *The Audit Society.* Oxford: Clarendon Press.

———. 2007. *Organised Uncertainty: Organising a World of Risk Management.* Oxford: Oxford University Press.

Rancière, Jacques. 2004a. *The Politics of Aesthetics.* London: Continuum.

———. 2004b. *The Philosopher and His Poor.* Durham, N.C.: Duke University Press.

Roux, Jacques, and Tierry Magnin. 2004. *La Condition de Fragilité: entre science des matériaux et sociologie.* St. Étienne, France: Publications de l'Université de Saint-Étienne.

Runciman, David. 2006. *The Politics of Good Intentions: History, Fear and Hypocrisy in the New World Order.* Princeton, N.J.: Princeton University Press.

Schaffer, Simon. 2003. "Enlightenment Brought Down to Earth." *History of Science* 41: 257–68.

———. 2005. "Public Experiments." In *Making Things Public,* ed. B. Latour and P. Weibel, 298–307. Cambridge, Mass.: MIT Press.

Shapin, Steven, and Simon Schaffer. 1985. *Leviathan and the Air Pump: Hobbes, Boyle and the Experimental Life.* Princeton, N.J.: Princeton University Press.

Simondon, Gilbert. 1992. "The Genesis of the Individual." In *Incorporations,* ed. J. Crary and S. Kwinter, 297–319. New York: Zone Books.

Stengers, Isabelle. 1997. *Power and Invention.* Minneapolis: University of Minnesota Press.

Stengers, Isabelle, and Bernadette Bensaude-Vincent. 2003. *100 Mots pour commencer à penser les Sciences*. Paris: Les Empêcheurs de Penser en Rond.

Strathern, Marilyn. 1995. *The Relation: Issues in Complexity and Scale*. Cambridge: Prickly Pear Press.

Tarde, Gabriel. 1999. *Monadologie et Sociologie*. Paris: Les Empêcheurs de Penser en Rond.

———. 2001. *Les Lois de L'Imitation*. Paris: Les Empêcheurs de Penser en Rond.

Thrift, Nigel. 2008. *Non-representational Theory: Space, Politics, Affect*. London: Routledge.

Waldron, Jeremy. 1999. *The Dignity of Legislation*. Cambridge: Cambridge University Press.

West, Harry, and Todd Sanders, eds. 2003. *Transparency and Conspiracy: Ethnographies of Suspicion in the New World Order*. Durham, N.C.: Duke University Press.

Whatmore, Sarah J. 2006. "Materialist Returns: Practicing Cultural Geography in and for a More-than-Human World." *Cultural Geographies* 13: 600–9.

Wheatley, Jonathan. 2005. *Georgia from National Awakening to Rose Revolution: Delayed Transition in Post-Soviet Politics*. Aldershot, U.K.: Ashgate.

Whitehead, Alfred N. 1985. *Science and the Modern World*. London: Free Association Books.

Žižek, Slavoj. 2004. Afterword to *The Politics of Aesthetics*, by Jacques Rancière. London: Continuum.

5 Plastic Materialities

GAY HAWKINS

YOU SEE IT WALKING INTO THE SUPERMARKET: an image of a plastic bag with a big black cross over it and the words SAY **NO** TO PLASTIC BAGS emblazoned above. The message is clear: bags are bad. How did it come to this? How did this flimsy, disposable thing acquire such a shocking reputation? How did using one in public come to mark the shopper as irresponsible? How did this humble object come to have such a claim on us?

As the supermarket poster shows, bags have changed. They have become contested matter: the focus of environmental education campaigns designed to demonize them and reform human practices. In this version of public pedagogy, there is no room for ambiguity about the meanings or affects of plastic materiality. As scientists discover marine life choking on bags and environmental activists document the bags' endless afterlife in landfills, plastic bags are transformed from innocuous, disposable containers to destructive matter. Say-no campaigns deploy a command morality designed to remind shoppers that bags are now problematic, yet another thing to register in the circuits of guilt and conscience that enfold us within forms of rule.

But what of the bag in all this? It appears as a passive object of reclassification. Scientific knowledge and social marketing frame it as bad stuff to be rejected by the environmentally responsible subject. But is this the only way in which plastic bags act on or make claims on us? If not, then we might wish to ask a different question: how does environmental education—and its command moralities—come to organize ethical transactions between plastic bags and humans in ways that disavow other transactions, other ways of encountering bags, that might suggest different, and more ecologically careful, modes of living? In many versions of environmental ethics, destructive matter manifests what Noel Castree (2003, 8) describes as a "materialist essentialism." It is seen as having clearly definable properties that are ontologically fixed. And as Castree (8) explains, "these properties can, in the final instance, be appealed to by environmental ethicists (explicitly or implicitly) to anchor claims about the who, what and how of ethical considerability." Despite the recognition of relational ontologies and calls for an ethics based on "transpersonal connections," the tendency is to demonize environmentally dangerous matter as materially irreducible and to fall back on the ontological distinctions that sustain this such as subjects–objects or nature–culture. This tendency inevitably privileges humans as the source of ethical awareness and action. Whereas natural matter is recognized as ethically significant and as a site of communicative vitality, destructive artificial material is afforded no capacity to affect us in ways that might call forth *other* ethical responses. Humans are not invited to be open to the affective intensities of plastic matter; rather, they are urged to enact their ethical will and eliminate it.

This is how ethics slides into moralism. As much as one may agree that the world would be a better place without plastic bags, the moral imperative to refuse them denies the complexity of contexts in which we encounter them and the diversity of responses bags generate. It fixes the material qualities of plastic bags and assumes to know the affects they trigger. Catastrophic images of plastic bags as pollutants link them to the end of nature and fuel a sense of disgust and horror. There is no possibility that plastic bags might move us or enchant us or invite simple gratitude for their mundane convenience. There is no sense that they might prompt us to behave differently. Instead, approaches to environmental ethics invoke essential characteristics that deny the contingency of ethical constituencies and relations. In doing so, they may also deny the affective dimensions of ethics and the ways in which corporeal interactions with the world are always mixed up with ethical reasoning and negotiations.

My goal is to get beyond this impasse—to examine *how* plastic bags come to matter without recourse to a materialist essentialism and without putting humans at the center of the story. By letting plastic bags have their say, I want to open up a different line of thinking about the relation between ethics, affect, and the environment, one that begins from the modest recognition of plastic bags not as phobic objects ruining nature but as things we are caught up with: things that are materialized or dematerialized through diverse habits and associations. By refusing to situate plastic bags in a moral framework, as always already bad, we might begin to see their materiality as more contingent and active. Bags may cease to be only ever passive and polluting, the source of dangerous environmental impacts, and may become instead participants in various everyday practices in which the materiality and meaning of *both* bodies and bags are fashioned. This is not to say that materiality is reducible to relations; rather, it is to suggest that different associations *make present* different material qualities and affects, and this contestability of matter is fundamentally implicated in ethical deliberations. The challenge is to understand the ways in which different plastic materialities become manifest and how these reverberate on bodies, habits, and ecological awareness.

In seeking to understand the potency of plastic bags in everyday conduct and political association, my concern is with the performativity of plastic: the ways in which its distinct materiality is realized in diverse arrangements. This focus on performativity, or "thing-power materialism," as Bennett (2004, 348) calls it, makes it possible to recognize the variety of plastic materialities beyond their framing as environmental hazards. For Bennett, *thing-power* means the specific kinds of materiality that are often obliterated by human habits of objectification and classification. In claiming that "there is an existence peculiar to a thing that is irreducible to the thing's imbrication with human subjectivity," Bennett (348) is not arguing for an essentialized materialism; rather, she is insisting that things have the capacity to assert themselves, that their anterior physicality, their free or aleatory movements, can capture humans as much as humans like to think they have the world of things under control. Recognizing the thingness of things is not to deny the dense web of connections in which they are always caught up. It is simply to be open to the powers of matter and the possibility that bags might suggest different ways of acting *with* them rather than *on* them.

Rather than begin from the environmentally aware subject committed to eliminating plastic bags in the name of nature, I want to investigate how plastic bags help produce this subjectivity, how they become involved in distinct forms of self-cultivation and ethical reasoning, and how the unpredictability of affect might be implicated in plastic bag ethics. This means paying close attention to plastic bag habits. For it is in the relational imprints of meaning back and forth between the body and its environs that habits, as practical techniques, become sedimented and confront us as a kind of second nature. Habits have a materializing power on both subjects and objects. They bind us to the world at the same time as they *blind* us to it. And this is the problem and the possibility of habits: when they break down, or when they are problematized, we are launched into new relations with the world. A moralized language of habit assumes that these new relations come from a virtuous subject who is responding to the appeal of reason and reforming his behavior. My claim is that it may well be the performance of plastic materiality, or the transformative character of affectivity, that disrupts habits and prompts new perceptions and ethical praxes.

To pursue these issues, I am going to use three plastic bags: the banned plastic bag of environmental education, the plastic bag as handy domestic container, and the dancing plastic bag from the hit Hollywood movie *American Beauty.* Each of these bags manifests distinct plastic materialities, and each can generate different affective energies. Though they share the same material qualities, the performance of these qualities in different assemblages is evidence that plastic materiality cannot be essentialized, and nor can ethics. Rather than being a set of fixed principles in the name of moral reason, these plastic bags reveal the fundamental porousness and instability of ethics. In these examples, ethics emerges as ubiquitous, affective, and thoroughly imbricated with corporeality. Acknowledging this ethical instability does not mean an abandonment of environmental politics but a different mode of political thinking, less concerned with dissensus and contestation and more concerned with speculative practices and improvisation. Each of these plastic bags "forces thought," to use Isabelle Stengers's (2005) phrase. They make themselves known in different ways, and in being open to these different knowledges, it may be possible to enlarge the politics of plastic bags—to imagine different modes of thinking, feeling, and acting with them.

Say No to Plastic Bags!

Campaigns to eliminate plastic bags have become a common fixture in countries where environmentalism is highly organized. Sometimes run by governments, sometimes by green or activist organizations, these campaigns focus on reducing plastic bag use by urging consumers to choose more sustainable alternatives. In Australia that alternative is, most often, a green shopping bag made out of long-lasting polypropelene with an environmental slogan on the side. In encouraging shoppers to voluntarily reject disposable plastic bags, say-no campaigns are explicitly pedagogic; their intent is to reform populations and change everyday habits. But how do they do this, and what is the role of the plastic bag in this process? By investigating how environmental campaigns problematize plastic bags and shopping practices, it is possible to see how these mundane objects become caught up in new associations that organize a distinct set of interfaces between bodily habits, materiality, and ethical reasoning, and how, in activating techniques of conscience, plastic bags participate in fashioning an environmentally concerned shopper.

Using a range of scientific information about environmental impacts, say-no campaigns frame plastic bags as hazardous, and in the same moment, they invite shoppers to engage in self-scrutiny and reflect on the everyday conduct around them. This framing is explicitly moral. It involves fixed oppositions, such as environmentally friendly–environmentally hazardous, and it appeals to categorical imperatives, such as protecting nature or global ecological survival. This is the larger scale in which minor habits and their impacts are situated. In constituting plastic bags as a "matter of concern," as Bruno Latour (2005) might say, say-no campaigns activate specific aspects of the materiality of the plastic bag: their slow process of decomposition, their tendency to trap or choke marine animals, their oppressive ubiquity, and so on. These material qualities are not representations or social constructions; rather, they are a particular aspect of plastic materiality that is made present to transform the meaning of the bag from innocuous container to polluting and recalcitrant matter. These reframings of the bag expose its material afterlife and extend the ethical imagination of the shopper. They reveal disposability as a myth and establish a network of connections and obligations between ordinary habits and the purity and otherness of nature. In this way the bag becomes capable of generating not only environmental concern but also guilt.

Guilt is a powerful reminder of the claims matter can make on us. Adopting new conduct that avoids plastic bags involves an acceptance of plastic materiality as dangerous and a willingness to change one's relationship to that matter out of a sense of obligation to the environment. This new network of relations between bags, shoppers, and nature involves practices of self-monitoring and self-discipline that Ian Hunter (1993, 128) describes as "techniques of conscience." The capacity of plastic bags to make some shoppers hesitate before they reach for one is only successful if subjects are receptive to the ethical obligations the bag's materiality poses to them, if they have a conscience.

According to Foucault (1985, 29–30), conscience is a product of a range of techniques of the self that have come to constitute distinct styles of subjectivity. To be a subject now means cultivating particular modes of reflexivity. It means developing special ethical techniques and capacities. These techniques and capacities are historically variable in their form and targets. Their presence is not evidence of a foundational interiority grounding the subject; rather, it is evidence of shifting regimes of living and self-cultivation. Techniques of conscience make the self into an object of ethical attention; they show how subjects problematize and modify their conduct on the basis of ethical principles to which they aspire. And, as say-no campaigns reveal, matter can play a key role in activating techniques of conscience. It can prompt changed practices that are justified by appeals to various moral codes and principles. Though environmental education campaigns, and their psychological logics, assume that ethical agency resides in the raised consciousness or "awareness" of the concerned individual, that individual is contextually situated, and those contexts involve multiple interactions with plastic materiality.

Public campaigns about the hazardous materiality of plastic bags are successful not simply because they have reeducated shoppers but because they have animated the materiality of bags in powerful ways. They have made the plastic bag an intermediary between an interior reception of an ethical command and the mobilization of the will to abide by it (Bennett 2001, 156).

Say-no campaigns run by governments or environmental nongovernmental organizations show how plastic bags have become implicated in processes of moral self-regulation and conscience; how circuits of guilt, self-reproach, and virtue have become enfolded with ordinary acts of shopping; and how, in activating techniques of conscience, the plastic bag participates in shaping an environmentally aware subject. The force of matter in this process, its capacity to prompt certain practices in particular arrangements, is evidence of the formation of a distinct ethical constituency in which changed interactions between bags and bodies produce new effects. These effects are more than just reduction in use; they also involve the formation of collectivities. For the shopper, recognition of bags' polluting materiality is a source of ethical concern and a prompt to reject them. When that shopper arrives at the supermarket checkout and presents her green ecobags, the absence of the plastic bag is a public declaration of environmental awareness. The ecobag as an accessory becomes a marker of a nascent political community of concerned subjects whose collective rejection of plastic bags implicitly links them. In the same way, the shopper struggling across the parking lot, arms weighed down with full plastic bags, is vulnerable to public scorn about his bad habits. How many times at the checkout have we heard a shopper declare guiltily, "Sorry, I forgot to bring my green bags"?

There is no question that say-no campaigns involve differential degrees of agency on the part of plastic materiality and that the ethical constituency formed by these campaigns is an environmentally aware subject who encounters the bag as hazardous matter. There is also no question that the affective energies that are generated by this style of environmental campaign involve various registers of moral righteousness and anxiety. However, as effective as these campaigns have been in some places in reducing the use of plastic bags and developing enhanced ecological awareness, their limits must also be acknowledged.

William Connolly (1999, 195) argues that conscience and other code-driven moral techniques are crude and blunt tools for coping with the world. Their tendency to ground moral or political action in law, God, global survival, consensus, or any other categorical imperative makes them blind to the ambiguous and disturbing aspects of many encounters. The moral weight of codes can too easily turn obligation into duty, guilt, and resentment: "I *should* do _____ because the environment is suffering, because I am law abiding, because I am virtuous." This is obligation working in the interests of human mastery and self-certainty, obligation that implicitly maintains the stability of being. Though say-no campaigns have only been successful because they have animated the materiality of bags and implicated humans in new relations with them, the differential agency of the bag in this process is disavowed. It is something to be controlled by human will, not a participant in an emergent ethical constituency. The logic of categorical imperatives and prohibition privileges the concerned and virtuous shopper as the source of ethical action and change. In this way, obligation and guilt suppress the capacity of the bag and deny the ways in which its materiality always exceeds moral framings.

A Sticky Plastic Bag

Consider another encounter with a plastic bag: a Monday-morning before-school panic that it is swimming training this afternoon but the bathing suit is wet. It cannot be put in a schoolbag because it will make books and lunch damp. You search under the kitchen sink for a plastic bag. They are hard to find in this house because reusable ecobags get the shopping home each week. But then you see one lurking in the dark recesses of the cupboard. You pull it out and hand it to the child: "Here, put your bathing suit in a plastic bag." The bag is grabbed, the child tries to open it, the plastic is sticky and slightly resistant to this gesture, and then its waterproofing and container possibilities are revealed and the wet bathing suit is stored safely. You feel gratitude for the humble practicality of the plastic bag.

In this ordinary moment, the bag does not simply perform utility; it also presents its materiality as something to be experienced and negotiated. The sticky plastic makes a polite request to the human to be patient and persistent, to rub her thumb and finger together to get a better grip. When the bag opens, panic is converted into appreciation. This is a collaborative process in which the meaning and materiality of the human and the bag shift. The bag's plastic presence is noticed not as a bad matter but as what John Law calls (2004, 84) "in-here enactment." For Law, this means the processes whereby material presence is enacted into being in distinct relations and practices. Presence is what is *made present* in particular relations. However, at the same time, it involves manifest absence because presence is always incomplete, always limited and contestable. The manifest absence in this encounter is the moralized plastic bag of environmental awareness and the virtuous identity of the ethical consumer. In this particular web of domestic associations, the in-here enactment of the bag generates experiential networks of obligation that disturb neat oppositions between environmentally aware subject and hated object. The plastic bag has become a player in a different process; in asserting its material presence, it disrupts knowledges of it as dangerous and destructive. Its mundane practicality challenges the circuits of guilt and conscience that drive command moralities: *say no to plastic bags!* Instead, the in-here enactment of the bag reveals a different plastic materiality that rearranges conduct and perceptions. Our response to the invitation from the bag to be patient disturbs arrogant senses of human agency and mastery. This inanimate thing is animate: it is suggesting particular actions.

In this familiar example, a different form of problematization is in play—we could call it *pragmatic problematization*. The bag is making itself known as sticky. As already outlined, say-no campaigns involve moral problematization. In seeking to connect ethical praxis to the survival of the planet, they have to fix the material qualities of bags as bad. The fact that bags always exceed this framing cannot be acknowledged because it introduces ambiguity and contingency into a politics driven by categorical imperatives and the logic of human intervention and prohibition. Whether the target of political change is shoppers or macroassemblages like retailers or state policy, the aim is to eliminate the problematic bag and save nature. In a world represented as drowning in plastic bags, a concern with *how* plastic materiality is performed in various associations seems both indulgent and grotesque. Yet this is precisely what the sticky plastic bag does. It does not problematize nature or bad habits; it simply makes us aware of how plastic materiality can be both resistant and useful. This plastic materiality does not have political capabilities, but it does have the capacity to render unstable moral certainties and their human centeredness and to suggest that there are multiple sites of agency in the world.

When the sticky plastic bag asserts itself, we are reminded of how enmeshed we are with it. This bag presents itself to us as a practical resource for being. We do things with it, leave our trace on it, but this does not mean that it is completely subordinate to human action. It has a life of its own that we have to accommodate in our activities. This bag puts questions of action and practice at the center of ontology, and what we can do with it becomes central to how we know it. As Elizabeth Grosz (2001, 168–69) says, "the thing poses questions to us, questions about our needs and desires, questions above all of action: the thing is our provocation to action and is itself a result of that action." This is how the sticky plastic bag suggests an alternative to a moral response to bags as bad stuff to be eliminated. By insisting that we work with it, the bag makes us aware of the ambiguity of intercorporeality and our complex entanglements with matter. The need for cooperation short-circuits guilt and makes us open to the thing-power of plastic materiality.

Pragmatic problematization suggests a different mode of political analysis. According to Collier, Lakoff, and Rabinow (2004, 3), it involves a shift from a first-order observer concerned with intervention and repair of a situation's discordancy to a second-order observer whose task is to see a situation "not only as a given but equally as a question." This approach to problematization offers a technique for understanding how, in a given situation, multiple constraints are at work, but also multiple responses. This resonates with Deleuze and Parnet's (1987) understanding of politics as a process of *active experimentation.* If discordant situations disturb and defamiliarize, if they make trouble for previous ways of understanding and acting, then they also create spaces of possibility where other ways of being may be revealed. In the shift from intervention to experimentation, the scale of politics is transformed: experimental practices are played out *between* large-scale macropolitical institutions and processes and the subinstitutional movements of affect, habit, and minor practices. Micropolitics occurs at the level of detail, demeanor, feeling, and response; it reveals the ways in which forms of embodiment and sensibility shape being in the world and how the world in turn reverberates on bodies. Central here are the qualitative dimensions of deviant minor practices and material habits and the ways in which they can disrupt normativity and moral codes. For it is precisely in these minor practices, like being responsive to the plastic bag suggesting that you be patient, that matter might shift perception and invite experiments with new practices. For Connolly (1999, 149), macro- and micropolitics do not exist in relations of opposition, and nor should they be ranked on a scale of importance. They are interconnected, and "politics becomes most intensive and most fateful at those junctures where micropolitics and macropolitics intersect."

American Beauty: A Dancing Plastic Bag

In turning now to a cinematic plastic bag, my aim is not to do an interpretation or reading of the scene. Though the institutional organization of cinema is central to allowing this plastic bag to be represented and to circulate, I am not concerned with the plastic bag as an object of signification, nor I am concerned with an ideological critique of the cultural messages of the film and the plastic bag's role in these. As with the banned bag on the say-no poster and the handy bag for a wet bathing suit, my interest is in the relations between performativity, materiality, and micropolitics. I want to think about how the bag works in this particular cinematic assemblage, the kinds of connections it establishes with the responding body, and the ways in which a distinct thing-power materialism is put into play. I am concerned with film technique, but not in the way this is normally understood. By technique I do not mean the formal processes of meaning making but rather the ways in which cinema as a complex cultural apparatus mediates materiality and can induce responses in the most visceral registers of the self.

Films can move, surprise, and disturb us in ways we barely understand until after the impulse has reverberated across our flesh. Their influences can work beyond the level of consciousness and thinking. While some might argue that this is precisely the problem with cinema, its capacity to manipulate and distort, I take my lead from Connolly (2002, 12–13), who argues that it is in these very moments when the body is captured by the screen, when habitual modes of viewing are disrupted or when the powerful forces of affect are mobilized, that cinematic technique can become implicated in the reorganization of perception and the dynamics of micropolitics. For Connolly, cinema is one of several key sites that reveal the complex relations between affect, thinking, and language. Its capacity to concentrate and intensify images, sounds, voice, and music can affirm the powers of the sensing body and the layered processes of thinking. In his schema, consciousness is reframed from a self-sufficient and disciplined zone of knowing and representing to a zone that is subject to massive layers of sensory material and filtering. The embodied work of thinking involves "powerful pressures to assimilate new things to old habits of perception" (164), and it is in the interstices of these pressures that new ideas and sensibilities might be created, that a politics of active experimentation might surface: "these new ideas, concepts, sensibilities and identities later become objects of knowledge and representation. Thinking is thus creative as well as representative, and its creativity is aided by the fact that the *process of thinking is not entirely controlled by the agents of thought*" (65; italics added). Connolly is describing the transformative power of affect here, and this analysis is invaluable for making sense of how plastic bags might become implicated in an expanded sense of politics.

The notorious plastic bag scene in *American Beauty* is both moving and disturbing. It triggers specific geographies of affect that engulf the viewer with the vivacity of an impression. In the structure of the narrative, this scene has the insistent singularity of arriving out of nowhere. It appears on the screen as part of Ricky's home video collection. He and Jane, the two teenage protagonists, are getting to know each other in his bedroom. Ricky offers to show Jane his favorite home video, what he considers to be "the most beautiful thing in the world." The screen is suddenly filled with a grainy digicam image of a plastic bag being blown about in the wind. The bag swoops and weaves; it is tossed about with the leaves in the gutter against a backdrop of a nondescript brick wall on an anonymous urban street. The setting is irrelevant; as the camera sticks to the bag, its mobility and vitality become the center of attention: the bag is dancing for us.

In terms of technique, the actual cinematic language of this scene is deliberately and self-consciously singular. The logic of this sequence is disruption: from slick, high-gloss Hollywood production values to a sudden digicam aesthetic with its documentary and avant-garde resonances. However, this opposition between documentary and fiction gets nowhere near explaining the scene's singularity. It cannot be reduced to an irruption of the real into the fictional, nor can it be explained by invoking referentiality or objectivity. An analysis driven by some kind of genre fundamentalism does not get far. As so much documentary and film theory has explained, fiction and nonfiction inhabit each other. The nature of that bag's capacity to surprise, to move the viewer, comes not from the shock of genre impurity, a little dash of documentary out of context, but from quite different forces—specifically, the way in which the scene captures the distinctive thing-power materiality of the plastic bag and the affective force field that is generated by this.

What much recent writing on affect insists is that affect is, in many senses, prior to feelings and emotions; having a feeling is not the same as knowing it is a feeling. Being able to name a feeling, to classify feelings within some kind of emotional taxonomy, is to render affects available to consciousness, to make them knowable, to recognize them. But we are in affect, participating, before this happens. Affect precedes these kinds of classificatory and cognitive activities. For Brian Massumi, the gap between affect and emotion is a protosubjectivity. Affects remind us of the body's intensities and multiplicities, of the autonomy of experience. They are a surplus, an excess; they are about those registers of the self that escape the knowable, manageable subject: "the unbiddenness of qualitative overspill" (Massumi 2000, 186).

What is valuable about Massumi's account of affect is the way it makes trouble for all those epistemologies that begin with a knowing subject ready to act on the world or be acted on. For the body in affect is not subjectivity to the world's objectivity; rather, it is a body in transition, a body in relation. To respond, to have a response, is to be in a relation. This is why Massumi argues so emphatically that affect is relationality. Drawing on the work of William James, he argues that relationality is already in the world, and to be in the world and to participate in it is to be in an ever-unfolding relation. Thinking about affect in this way means an abandonment of the subject–object dualism. What is needed instead, according to Massumi, is a notion of continuity and discontinuity that is not framed in terms of opposition but as a *processual rhythm*. This opens up an understanding of how we are in and of the world, how being is a kind of ontological tension between manipulable objectivity (reality and all those things that represent it, e.g., language and documentaries) and elusive qualitative activity: all those things that break in from the outside, that surprise, that enliven, that introduce unpredictability—a dancing plastic bag, for example (Massumi 2000).

Massumi's idea of continuity–discontinuity as an ongoing processual rhythm shows that it is not so much a question of the generic shock of a documentary moment rupturing the diegetic space of a slick Hollywood narrative—though the bag scene obviously does this—but is more a question of *how* the bag performs. It is not just that it is real but that it is more than real; it is alive and animated. The bag reveals the extraordinary vitality of plastic materiality: its capacity to respond to the movement of wind, its lightness and transparency, its shape-shifting flexibility. And this energy and beauty generate an unexpected and intense affect that registers somatically, beneath and before consciousness. This scene triggers a different rhythm in the viewing body. To think about its affective force is to think about a discontinuity in perception—not the dualism of a cinematic text and a spectator, but an unpredictable, uncontained multiplicity, in which the viewer is caught up with or responding to the materiality of an enchanted plastic bag. Body *and* bag open to each other, crossing over and exchanging various materialities.

The dancing plastic bag disturbs habituated modes of viewing and existing meanings. It challenges a framing of plastic materiality as inert rubbish. This bag does not just move us; it forces us to notice its materiality. Though it may feel like we are seeing a plastic bag as if for the first time, as authentic matter, as the pure thing itself, this materiality is still a product of distinct assemblage, from the technologies of cinema to the perceptual trainings of the audience to everyday waste habits. This is not to reduce the plastic bag to an effect of these technologies, a mere representation; rather, it is to claim that the thing-power materialism in play is both real and contingent, an outcome of diverse techniques that seek to render manifest the force of matter and frame it in distinct ways.

In *American Beauty* the plastic bag does not have any narrative function; the movie would do fine without it. It appears to be there just to generate affect. But how does it do this? We cannot explain this exclusively by the logic of the referent; bags are not usually that moving. There is no question that this scene is an image of redemption, hated or ignored matter rendered beautiful and alive, but that is not really the issue; rather, it is what cinematic techniques do to things that is more important. As Lesely Stern (2001) argues, the ways in which things acquire meaning and affect in the cinema have to do with how they are captured by the camera, their mutability, and their implication in the quotidian—the social life of things before they are framed by the cinematic gaze.

Stern's argument makes trouble for analyses of documentary as a genre that captures the real; rather, she insists on the need to theorize cinema's relation to the quotidian, the different ways in which cinema captures or frames everyday matter. She posits a tension between two specific cinematic techniques, the histrionic and the quotidian, which exist in a generative tension with each other. Cinema has always been about ordinary things; it is what it does with them that is important, how it conjures them and invites them to perform. In *American Beauty,* there is no question that the plastic bag is performing for the camera and that, in this long, evocative sequence, while the digicam stays on the bag, the audience experiences a suspension of narrative, a diversion. Using Stern's argument, we can see how, in this diversion, narrativity and histrionic techniques are suspended in deference to the quotidian, to a performative technique that involves a gestural framing of the thing, highlighting it and watching it move. We have a sense of the "world caught off guard, unposed, real" (Stern 2001, 327). As Stern says (327), "editing is deflated in deference to the primacy of the real allowing a kind of minimal inflation of real time." This discontinuity creates a new rhythm, where a different relation between temporality and affectivity is established, one that is in the realm of emotional duration.

Cinema's fascination with the performance of things is continually highlighted in *American Beauty* with references to camcorder culture. Ricky uses his digicam to try to capture the ineffable singularity of matter. He is not using the camera to mediate the world, to separate himself from it; rather, it is a tool for trying to simultaneously render the material force of objects and convey the insistent autonomy of his experience of them. Or to use another term, it is a tool to capture affect as *relationality*, as a sense of being caught up in the world rather than the center of it. Consider his narration of the plastic bag scene:

> It was one of those days when it's a minute away from snowing. And there's this electricity in the air, you can almost hear it, right? And this bag was just . . . dancing with me. Like a little kid begging me to play with it. For fifteen minutes. That's the day I realized there was this entire life behind things, and this incredibly benevolent force that wanted me to know there was no reason to be afraid. Ever. . . . Video's a poor excuse, I know. But it helps me remember.

The technical possibilities of the portable digicam dramatically foreground cinematic performativity: what a camera does to things, how it conjures them up in ways that make them seem enchanted. But at the same time, things have a life of their own. As Ricky says, "video's a poor excuse."

The affective force of this scene comes from the glimpse it offers of that quotidian world of plastic bags existing not as passive objects waiting to be refused but as elusive actants: things that have the power to capture us in new relations. And this affective force is potentially transformative. To be moved by a bag, to feel enfolded with its materiality, to sense a force field of unexpected energies surface as it performs on screen, is to be caught up in completely different habits of perceiving and responding. This surprising encounter undoes the circuits of guilt and obligation through which subjects feel the pull of prescriptive moral codes. This is not a bag to be refused in the interest of producing a virtuous identity but rather a bag that is generating an intense entanglement, a shock of new feelings for plastic matter and what it can do. This bag is not appealing to conscience.[1] This dancing plastic bag flattens out the subject–object hierarchy, generating an assemblage of affective connections and reverberations in which thing-power materialism, "the recalcitrance or moment of vitality of things" (Bennett 2004, 354), is put into play. For Bennett the affective agency of thing-power is inextricably linked to ecological awareness:

> Thing-power is a kind of agency, [it] is the property of an assemblage. Thing-power materialism is a (necessarily speculative) onto-theory that presumes matter has an inclination to make connections and form networks with varying degrees of stability. Here, then, is an affinity between thing-power materialism and ecological thinking: both advocate the cultivation of an enhanced sense of the extent to which all things are spun together in a dense web, and both warn of the self-destructive character of human actions that are reckless with regard to other nodes of the web. (Bennett 2004, 354)

The Politics of Plastic Bags

The concern in this chapter has been to understand how plastic materiality is made present in various arrangements and to investigate how this materiality exhibits differential degrees of agency. By letting plastic bags have their say, the aim has been to challenge the tendency of environmental ethics to material essentialism and its blindness to the shifting processes whereby bad stuff becomes ethically and politically significant. Deploying a more-than-human mode of inquiry makes it possible to see how plastic bags are implicated in both fashioning and *disturbing* an environmentally concerned subjectivity and how these disturbances and problematizations are implicated in an expanded politics of plastic bags.

The capacity of plastic bags to disturb is a product of shifting associations and interactions. While environmental awareness campaigns such as say-no campaigns have been crucial in animating the disastrous effects of plastic materiality, they involve a limited style of address and politics. Their technique of moral problematization makes the bag an intermediary in the relays between the interior reception of an ethical command and the mobilization of human will to abide by it. However, in activating techniques of conscience, these bags explicitly fix political and moral concepts. They are always already bad, and they amplify the mastery of humans as the source of political change and environmental survival: *say no!*

The handy plastic bag generates a very different network of relations. Its mundane practicality and its capacity to make humans be patient is disturbing. This thing-power materialism reveals the fundamental porousness and instability of command moralities. It generates a pragmatic problematization of the bag that has the potential to trouble and deconstruct the politics of environmental education. In reassembling how bodies experience the bag and behave around them, a different micropolitics of habit, sensibility, and perception is activated. These micropolitical responses are mixed up with code-driven disciplines. They reveal another layer of response flowing through the certainties of ethos and command. The handy plastic bag and the dancing plastic bag are powerful evidence that circulating through every dutiful or correct practice are moments of responsiveness and affect that capture subjects in new relations to the world. These are bags that prompt different feelings and knowledges and that enlarge the reach of politics beyond public reason and conflict into the terrain of the visceral, the situational, and the experimental.

Connolly argues that disturbance and unexpected responses are fundamental to politics because they show how affect is implicated in the composition of new sensibilities and associations. For Connolly, responsiveness is a condition of possibility; it opens lines of mobility and difference within the self, and it is something that can be cultivated. An "ethos of critical responsiveness" (Connolly 1999,

69), to use his term, connects political experimentation to affect and various practices of self-modification. It involves work on the self in the interest of recognizing the plurivocity of being *and* matter. Critical responsiveness decenters the human as the sovereign source of agency and change; in recognizing multiple sites of agency at play in the world, it invites an expanded politics attentive to how the force of matter might participate in generating new associations and ethics.

Notes

1 This section draws from my book *The Ethics of Waste* (Hawkins 2006, 36–39).

References

Bennett, Jane. 2001. *The Enchantment of Modern Life.* Princeton, N.J.: Princeton University Press.

———. 2004. "The Force of Things: Steps toward an Ecology of Matter." *Political Theory* 32: 347–72.

Castree, Noel. 2003. "A Post-environmental Ethics." *Ethics, Place, and Environment* 6, no. 1: 3–12.

Collier, Stephen, Andrew Lakoff, and Paul Rabinow. 2004. "Biosecurity: Towards an Anthropology of the Contemporary." *Anthropology Today* 20, no. 5: 3–7.

Connolly, William. 1999. *Why I Am Not a Secularist.* Minneapolis: University of Minnesota Press.

———. 2002. *Neuropolitics: Thinking, Culture, Speed.* Minneapolis: University of Minnesota Press.

Deleuze, Gilles, and Claire Parnet. 1987. *Dialogues II.* Trans. H. Tomlinson and B. Habberjam. London: Continuum.

Foucault, Michel. 1985. *The Use of Pleasure.* Trans. R. Hurley. New York: Vintage.

Grosz, Elisabeth. 2001. *Architecture from the Outside.* Cambridge, Mass.: MIT Press.

Hawkins, Guy. 2006. *The Ethics of Waste.* Lanham, Md.: Rowman and Littlefield.

Hunter, Ian. 1993. "Subjectivity and Government." *Economy and Society* 22, no. 1: 123–34.

Latour, Bruno. 2005. "From Realpolitik to Dingpolitik: or How to Make Things Public." In *Making Things Public: Atmospheres of Democracy,* ed. B. Latour and P. Weibel, 14–41. Cambridge, Mass.: ZKM and MIT Press.

Law, John. 2004. *After Method.* London: Routledge.

Massumi, Brian. 2000. "Too Blue: Colour Patch for an Expanded Empiricism." *Cultural Studies* 14, no. 2: 177–226.

Stengers, Isabelle. 2005. "The Cosmopolitical Proposal." In *Making Things Public: Atmospheres of Democracy,* ed. B. Latour and P. Weibel, 994–1003. Cambridge, Mass.: ZKM and MIT Press.

Stern, Lesely. 2001. "Paths that Wind through the Thicket of Things." *Critical Inquiry* 28 (Autumn): 317–54.

6 Halos: Making More Room in the World for New Political Orders

NIGEL THRIFT

THIS CHAPTER REPRESENTS one small part of a more general attempt to struggle over the hill of various Western philosophies, social sciences, and forms of politics to see a new, more open vista, one in which, through the articulation of an ontology of achievement, different associations are able to be made and made manifest, different togethernesses are thereby able to be forged, and different landscapes of possibility are subsequently able to be uncovered.

To limit what is clearly an enormous number of lines of enquiry, I have therefore fixed on just one aspect of this attempt, namely, the politics of the imagination. However, I make no apologies for choosing this topic as my touchstone, for different imaginative dispositions and propensities might well be thought of as the basis of political-moral authority.[1] For example, it can be argued that if power means the capacity to make somebody do what she would not otherwise do, then whoever possesses the capacity to influence others' imaginations has a good deal of power. To put it another way, political power is not only about controlling the means of coercion but also about controlling the means of imagination, where imagination is understood as the ability to express possible/play/pretend beliefs[2] and emotions that might become the basis of a better world. Equally, it is about working on so-called imaginative resistance, the possible beliefs and emotions that we resist imagining and accepting—either because we cannot imagine a possibility or because we do not want to imagine it (Nichols 2006).

Most particularly, I want to express the significance of this political project through one particular unifying visual device—the halo—which stands for a *change in the nature of the spatial representation*[3] *of imagination,* through work on the way in which emancipation and attachment are jointly framed, on the grounds that a good part of the political consists of the establishment of an effective imaginary (Castoriadis 1997).

I will introduce the conceit of the halo in the first part of the chapter. The three succeeding sections of the chapter then proceed to develop an argument through three different uses of this general framing device: as a means of approaching affective imitation and its ramifications, as a means of understanding the generation of semiotic intensity and thereby affective traps, and as the construction of forms of community that attempt to generate affective affinity in new ways. In each case, my purposes are political. Respectively, they are to displace prevalent models of political activism, to understand new forms of inhabitation and their possibilities, and to generate new locatives.

Halos

I have had three interrelated goals in mind in writing this chapter. To begin with, I want to trace out some of the changes in the political contours of our time, but I will do so by understanding "social" process as a mass of material entanglements slowly changing within daily practices, often without intentionality: "the mass seems to move with a life of its own. But the movement is built from the little micro-details of life" (Hodder 2006, 258). Specifically, I will be tracing out, perhaps in accelerated form, the further history of entanglement with material objects, a history in which these objects come to have greater and greater purchase on our lives, not only as the ability to construct continuity and to initiate change but also as small things too easily forgotten (Deetz 1996). I also have a second goal in mind, and that is pointing to certain ways in which Western academic thinking is fracturing, with interesting consequences. It is fracturing because it is realigning the roles of concepts, percepts, and affects, reworking space and time and taking on new partners (animals, materials). The result is a different take on the human, society, and imagination that is difficult to disentangle but can perhaps be made sense of as a generator of different kinds of "radical" politics (Tønder and Thomassen 2005). The third goal arises out of the previous two. I want to think about the possibilities of new creatures, understood as new compounds of life that act in unforeseen ways, *and* the new spaces in which such life can flourish, spaces that provide new lures to feeling, new powers to force thinking and invention, new schemes of purposefulness—or purposelessness—that can provide different means of moving us/them. In particular, I want to think about the kinds of model of affective agency that might be possible and, simultaneously (since they cannot be understood apart), how they might be fostered by speculative spaces that call to, provoke, and invoke these agents. In other words, I want to talk about new forms of intelligibility.

I will try to achieve these three general goals by using the conceit of the halo—standing, in general, as a means of beginning the process of considering and constructing new imaginative *plausibilities.* I will use the framing device of the halo because it conjures up an image which can have several kinds of grip. If I was to frame my argument in quasi-religious terms, then what is being sought through the agency of the halo is a device that will not so much unite as bring into correspondence that which is different without trespassing on that difference, and without trying to reduce what is puzzling to a predictable encounter; rather, because "each party may entertain its own version of the agreement" (chapter 1), *the art is in the achievement itself.* To repeat my opening remarks, the conceit of the halo is used to open up three specific dimensions of this art of achievement,[4] namely, the emergence and nurturing of infectious relationships, the design of spaces of inhabitation as both semiotic intensities and affective traps (Gell 1998), and the construction of new kinds of locative community.

The First Halo: Picturing Affective Contagion

In its most familiar guise, the halo is a staple of Christian religious iconography. Yet the halo is pagan in origin. Many centuries before Christ, it is thought that various peoples of the Mediterranean decorated their heads with a crown of feathers (Fisher 1995). "They did so to symbolize their relationship with the sun-god: their own 'halo' of feathers representing the fan of beams splaying out from the shining divinity in the sky. Indeed, people came to believe that by adopting such a 'nimbus' men turned into a kind of sun themselves and into a divine being" (16). Various pharaohs and emperors followed suit. Later, the halo appeared in the art culture of ancient Greece and Rome,[5] before being incorporated into Christian art sometime during the fourth century, adorning first Christ, later angels, and eventually saints.[6] Subsequently, the halo has had a rich history as the aureole that appears to emanate from beings of particularly intense spirituality, a history with its own shifts in representation—for example, during the Renaissance, when rigorous perspective became essential, the halo changed from an aura surrounding the head to a tilting disk that appeared in perspective, floating above the heads of saints, and then to a thin ring of light.[7] In later work, haloes would often appear by allusion or insinuation—as a circular pattern that falls behind a head or as an arc of a doorway.

In this chapter, I want to understand the halo, first of all, as signifying the construction of new forms of empathy that are simultaneously acts of identification with the feelings, thoughts, or attitudes of another and the imaginative ascription to a natural object or a work of art of feelings or attitudes considered to be present in oneself. The reason is that I think that something quite interesting is happening in Western thinking of late. It is, I believe, a result of the joining of certain strands of thought as a result of more general changes taking place in the nature of the apprehension of space, thus pointing the way to a new kind of political settlement, one that might allow a different kind of spiritedness to emanate, one based on an ethos of *craftsmanship of the moment* that can produce "instant" affective communities. In making this claim, I therefore want the halo to stand for an affective ambition that is the achievement of an *infectious relationship*.

To stake this particular claim, I want to fix on the haloes to be found in the works of one the very finest orchestrators of glances, gazes, and stares, namely, Giotto di Bondone (1267–1337). I will start by examining one of Giotto's remarkable frescoes, *The Meeting at the Golden Gate*.[8] In this fresco, the aging Joachim and Anna, Mary's mother and father to be, look each other straight in the eye in an atmosphere of solidity and stability. The halo that unites them is an expression of this atmosphere of happy encounter. Nothing could be more different in another of Giotto's finest frescoes, *The Betrayal of Christ*. There only Jesus and Peter have haloes. Judas does not: he remains a lonely and inner-directed subject, cut off from the affective flow. But I do not want to draw the obvious conclusions here about Euro-American Cartesian subjectivities and the like; rather, I want to fix on the face and how it is figured in these and other representations.

The uncanny semaphore of *the face* is a crucial element in both paintings, and it will be an important recurring motif in this chapter. After all, "the living face is the most important and mysterious surface we deal with. . . . Babies just nine minutes old who have never seen a human countenance, prefer a face pattern to a blank or scrambled one" (McNeill 1998, 4). Almost from birth, the gaze is fixed on the face, especially the eyes, as the baby constructs joint attention and intentional understanding (Eilan et al. 2005). In other words, the face, like language, is an aspect of public thought.

That fact can be illustrated in three ways. First, faces are one of the chief means by which affect is generated in the world. Their 46 separate muscles, the eyes, the mouth, the nose, and the ears allow a range of expression that is without peer in the natural world, and they produce or certainly enable many of the characteristics that are most notably human. Second, there is a history of the representation of the face. Certain facial states come to be increasingly represented over the course of history. For example, the smile figures more and more as a result of the increasing portrayal of the open mouth from the eighteenth century onward because of changes in social attitude—and better dentistry (Jones 2000)! Third, there is a technical history of the face, perhaps best illustrated by the history of cinema and its effects on our perception. For example, the close-up is a crucial way station in the history of the modern face, providing new means of attending to the face and new possibilities for relation, not least those arising out of the close-up's peculiar ability to generate both intimacy and threat, not least as a disembodied affect. The face itself becomes a frame, but it can also be located outside the subject in the world of technically assembled images (Hansen 2004). What seems evident is that the face is a crucial element of politics and the political. It was always thus, one might say. But the modern media have extended the range of body language in ways hitherto unforeseen, most especially by providing a set of stock affective scripts for which the face provides both the template and the chief means of operating.[9]

Most important, of course, the face is our chief means for producing and scripting affective effects. Through its medium, we exercise the capacity for mind reading that probably does most to distinguish us from animals.[10] Other creatures undoubtedly have pains, expectations, and emotions, but having a mental state and representing another individual as having such a state is a second-order phenomenon that, so far as we can tell, other creatures do not have or have in an attenuated form (Grandin 2005; Hurley and Nudds 2006). Currently the favored explanation for mind reading is the so-called simulation explanation, which effectively argues that

> people fix their targets' mental states by trying to replicate or emulate them. It says that mindreading includes a crucial role for putting oneself in others' shoes. It may even be part of the brain's design to generate mental states that match, or resonate with, states of people one is observing. Thus mindreading is an extended form of empathy. (Goldman 2006, 4)

In turn, the phenomenon of mind reading points to the importance of what I have called the infectious relationship, founded in the production of chains of imitation. This is a phenomenon that was noted early in the history of philosophy and psychology. For example, both Hume (1739) and Smith (1759) detected it in their writings on sympathy. In particular, understanding infectious relationships means understanding affective contagion, a central concern of turn-of-the-nineteenth-century social science in the form of the study of imitation and suggestibility. Imitation and suggestibility took shape as particular kinds of object through a hypnotic paradigm that was worked out through an interest in particular forms of psychopathology (such as hallucinations and delusions), even an interest in spiritualist forms of communication (Blum 2007). Imitation and suggestibility were sites for exploring all manner of issues such as consciousness, memory, personality, and communication. In particular they signified a "taking over" of the subject that defied normal economies of subject–object relations. However, subsequently, a move to psychoanalytic models of desire or to more discursive approaches to subjectivity ruled imitation and suggestibility out of court, and they fell into disrepair as a way of approaching social structuring.

But of late, imitation and suggestibility have been making a return, boosted especially by the rediscovery of the work of Gabriel Tarde on a somnambulist society and more general work on the construction of collective intelligence. Within cultural theory, viral models of contagion have been posited as explaining the workings of a range of phenomena, including ideology, governance, self-cultivation, and even resistance, but often in highly speculative ways that posit a kind of performative energetics without specifying what the source, content, or form of that energy might be.

What do we know about affective contagion (Thrift 2007)? To begin with, it means understanding affect as in large part a biological phenomenon, involving embodiment[11] in its many incarnations, but a phenomenon that is not easily captured via specular–theatrical theories of representation (Brennan 2004; Gumbrecht 2004). It brings together a mix of a hormonal flux, body language, shared rhythms, and other forms of entrainment (Parkes and Thrift 1979, 1980) to produce an encounter between the body (understood in a broad sense) and the particular event. Thus affective contagion is best understood as a set of flows moving in a semiconscious fashion through the bodies of human and other beings, not least because bodies are not primarily centered repositories of knowledge—originators—but rather receivers and transmitters, ceaselessly moving messages of various kinds on; the human being is primarily "a receiver and interpreter of feelings, affects, attentive energy" (Brennan 2004, 87).

This understanding points in turn to one more important aspect of affective contagion, namely, the importance of space, understood as a series of conditioning environments that both prime and "cook" affect. Such environments depend on prediscursive ways of proceeding that both produce and allow changes in bodily state to occur (Thrift 2006). Changes in bodily state require understanding that essentially autonomic hormonal and muscular reactions are continually transferring between people (and things) in ways that are often difficult to track. At the same time, they challenge the idea that the body is a fixed component of humanity. Humans might be more accurately likened to schools of fish briefly stabilized by particular spaces, temporary solidifications of affective pulses, most especially as devices like books, screens, and the Internet act as new kinds of neural pathways, transmitting faces and stances and providing myriad opportunities to forge new reflexes.

Thus concentrating on infectious relationships requires a cartographic imagination to map out the movement between corporeal states of being that is simultaneously a change in connectivity. But only a very limited range of spatial models currently exist that can understand flows of imitation/suggestion. Alongside familiar cartographic motifs from diffusion studies, these include certain very general metaphors that have arisen from the recent emphasis in social theory on mobility, a range of ways of staging space to conduct affect that can be found in performance studies, a set of artistic experiments with sites of affective imitation, often using the possibilities of modern electronic media and various kinds of conversation maps (Abrams and Hall 2006). However, it is also clear that certain technological advances, and especially those to do with mobile telephony and the Web, are making it easier to visualize flows of imitation, not least because they are themselves prime conductors.

Now we can also add in what we currently know about imitation and suggestibility. For imitation has become a paramount concern of the contemporary cognitive sciences, and this work is worth exploring in a little more detail because it contains many insights. In particular, imitation is now understood as a higher-level cognitive function, mirroring both the means and ends of action, and highly dependent on the empathy generated in an intersubjective information space that supports automatic identifications. For example, just as Hume and Smith might have predicted, hearing an expression of anger increases the activation of muscles used to express anger in others, especially those muscles found in the face. There is, in fact, only a delicate separation between one's own mental life and that of another, so that affective contagion is the norm, not an outlier. What differs between different cultures is rather what is regarded as the result of agency. Thus, for Western cultures, it can be a painful realization to understand how little of our thinking and emotions can in any way be ascribed as "ours"; it is very hard for Westerners to accept that broad imitative tendencies apply to themselves, both because they are unconscious and automatic and because the preponderance of apparently external influences threatens the prevailing model of an agent as being in conscious control of himself.

At the same time, it is important to stress that imitation is more than mere emulation. Imitation is different from simple emulation in that it depends on an enhanced capacity for anticipation, so-called mind reading (Thrift 2005). In particular, much of human beings' capacity for mind reading (whether this be characterized as inference or simulation) develops over years of interaction between infants and their environments and involves processing the other as "like me," and the consequent construction of high-level hypotheses like deception; that is, it involves a form of grasping that is innately physical and nonrepresentational because our privileged access is to the world, not to our own minds.

What seems clear, then, is that human beings have a default capacity to imitate, automatically and unconsciously, in ways that their deliberate pursuit of goals can override but not explain. In other words, most of the time, they do not even know they are imitating. Yet at the same time, this is not just motivational inertness. It involves, for example, mechanisms of inhibition, many of which are cultural.[12]

So it is that imitation generates a spectrum of affective states and most especially empathy, not only because the self–other divide can be seen to be remarkably porous but also because across it constantly flow all kinds of emotional signals. But this is a kinetic empathy, of the kind often pointed to in dance, a kinaesthetic awareness/imitation that is both the means by which the body experiences itself kinaesthetically and also apprehends other bodies (Foster 2002).

Having considered the infectious relationship through the medium of the face, I also want to use Giotto's Christian iconography of Joachim and Anna's gentle gaze to reflect on the possibility of forming new kinds of activists who are not the militant, even martial activists we have too often lighted on: those who are "self-confident and free of worry, capable of vigorous, wilful activity" (Walzer 1988, 313). In particular, I want to get away from the remains of the model of what Benasayag calls the "sad activist," always intent on configuring a center from which to think radical practices (Collectivo Situationes 2005), a model that puts so many off—not just the committed but also the uncommitted, for whom it can often appear that activists are "know-it-alls" (Eliasoph 1998). In other words, I want to talk about how it might be possible to face up to the world by generating new models of the activist that are not like Walzer's constant hero, strong of mind and will, and that may well be more effective in their desire to generate affective affinities that are open-ended: emergent capacities to empower rather than fixed programs that can be handed down.

Most particularly, I want to think about generating new moral-political stances (using this word to point toward the political and the spatial as aspects of each other) that express a different model of the political subject, stances that blur the boundary between mover and moved[13] that is so crucial to prevailing models of active agency. Among other sorties, the feminist literature (especially the diverse literature on feminine nature) and various anthropological investigations into fractal modes of being have both been attempting this reengineering of the subject over many years. Such sorties raise intriguing questions about many things we hold dear. Specifically, does passivity have virtues as a means of seeking political change?

In the midst of current world events, this question will no doubt sound discordant, but passivity, or so it seems to me, points to a different way of doing things, one that dates from early modern times and relies on a very different model of agency and a very different rhetoric of passions, both of which are dependent on understanding subjects as transmitters and receivers of infectious relationships. So far as the model of agency is concerned, it is crucial to understand that for "new creatures,"

> Agency admits of more positions than "autonomous agent." . . . In addition to the autonomous agent undermined by recent discourses, an "agent" can also refer to one who acts for another. . . . This

> deputized "agent" is not a "sovereign ruler" but a subject licensed by another authority to perform predetermined actions. The gap between "agent" and "autonomous agent" is crucial to seventeenth-century writers, who often deny "autonomy" but insist on "agency," both descriptively (each individual has agency) and prescriptively (all individuals must act in the world). As "agents" or "instruments" of another, individuals are simultaneously "acted by another," in Thomas Hooker's phrase, and enabled to act in the world. "Acted upon, we act," summarizes John Cotton. These writers desire agency only insofar as it differs from autonomy: they desire not "shaping power" over their identities and actions but to be shaped by another power. (Gordon 2002, 23)

So far as the rhetoric of passions goes, what is important in becoming a new creature is the mobilization of passions like pride and humility politically, "with the apparently 'active' vice of pride condemned for its ineffectiveness and the 'passive' virtue of humility serving the most dramatic revolutionary ends" (Gross 2006, 110). The religious model of a radical that was prevalent in the early modern period was connected to the practice of a feminized humility: the agent was an instrument, "the product of humiliation, anxiety, and soulful, feminine passivity, in the best sense of the word" (Gross 2006, 93), an agent "humiliated for collective sins past and reformed for the time to come": this is a "feminine" passivity, but not, I hasten to add, in any pejorative sense. It is fragility as a precondition of grace, passivity as a precondition of change.

What difference might this make? I am not sure. But take one classic example of a moral–political code in which the model of a constant, militant hero currently holds sway: courage and bravery.[14] Yet such a model varies markedly from what is considered seemly in other cultures. Our model of courage and bravery has its genesis in Greek notions of character. For example, for Aristotle, for every character trait, there is a vice of excess and a vice of deficiency:

> Aristotle says that true excellences of character—what are called the virtues—have in common that they strike the mean between excess and defect. Given a particular life-challenge, a courageous person

> will act in a way that avoids the excess of foolhardy recklessness, on the one hand, but also the defect of cowardliness, on the other. The courageous person will in any given circumstances, be able to find an appropriate way to behave courageously. That is what it is to strike the mean: to find an appropriate way to behave in circumstances in which it is possible to do too much or too little. (Lear 2006, 17)

In other words, bravery is a virtue falling somewhere between rashness (bravery in excess) and cowardice (bravery in deficit).[15] But what constitutes bravery and what constitutes too much or too little of it varies massively across cultures and through time. Since for much of the time, bravery and courage are clearly forms of "thinking without words" (Bermúdez 2003), depending very much on taking a stance to a situation, they rely on material symbols arranged as determinate spatial patterns for articulation. In other words, bravery and courage tend to be carried in particular material cultures that both illustrate and compound particular convictions: materials matter. I want to highlight two particular, contrasting examples of material cultures of bravery and courage, in which these virtues are thought through and as particular exceptional things/spaces, to make this point (Henare, Holbraad, and Wastell 2007).[16]

For the warlike North American Crow, to take one instance, bravery was materialized in the, to us, excessive practice of counting coup by planting coup sticks, that is, tapping an enemy with a coup stick before killing him,[17] and then counting the coup in a ceremony after the incident in the form of a feather for each incident, which could be worn in the hair or on a shield. But this was a particular kind of bravery:

> Obviously, the practice of counting coups valorized bravery—a trait that was necessary for the Crow to survive. Honor was accorded to the brave men, along with access to women, extra food, and other material benefits. Imaginative-desiring-erotic-honor-seeking-life was organized around this kind of bravery. . . . If the survival of the Crow tribe as a social unit had been the primary good, one might expect that the highest honor would go to the warrior who killed the first enemy in battle, or the warrior who killed the most. But to

> count coup it was crucial that, at least for the moment, one avoided killing the enemy. There is a certain symbolic excess in counting coups. One needed not only to destroy the enemy; it was crucial that the enemy recognize that he was about to be destroyed. (Lear 2006, 15–16)

Take another instance. But here, the example is passive and arises out of a long tradition of nonviolence, or, more accurately perhaps, the bravery of *hesitation*. I am thinking here, in particular, of standards of excellence lived by and instantiated in the Quakers:

> If hesitation gathers practitioners, it is because rules and norms are discursive expressions tentatively formulating something that has no definitive, authoritative formulation and hence does not communicate through obedience—which I call "obligations." Obligations communicate with the possibility of their betrayal. If ever a practice exhibited this possibility, it is that of the Quakers, who, as we know, did not quake in front of their God but in front of the menace of silencing what was asked of them in a particular situation, answering it in terms of preset beliefs and convictions. (chapter 1)

Just like the Crow culture of bravery, so the Quaker culture of bravery is instituted by a material culture: the material interface is the meeting house, which affirms the value of hesitation through the construction of an absolutely democratic space. In particular, the early North American meeting houses were built to a circular plan, thus producing an egalitarian acoustic in which everyone could be heard with the same volume wherever he spoke in the building (Rath 2003).

These are radically different, even opposed examples. But in combination with the previous discussion of passivity, they lead to some intriguing questions. What should we count as bravery? Is there a political economy of bravery? Might it be possible to reengineer bravery toward a "passive" affective model again, so that many more exemplary acts could be included?[18] What kind of material culture would be able to achieve this? What does bravery look like in a conglomerate of relationships that includes all manner of material and animal correspondents? Whatever the answer might be, *space* will be key, and so it is to space that I now turn, and especially to the multifarious spaces being formed by various forms of contemporary information technology, and most especially to the spaces being formed by ubiquitous-ambient-pervasive-persistent interfaces of various kinds, interfaces with which it is possible to have unconscious or semiconscious relationships (Bickmore and Picard 2005).

The Second Halo: Engineering Affective Environments

Nowadays, the word *halo* means as much to a Western audience as a best-selling series of computer games and associated comics[19] and graphic novels,[20] with a fanatical—and I do mean fanatical—fan base. On the basis of the old science fiction conceit of humans versus aliens on a halolike ring world, the Halo series first appeared in 2001, very much associated with the cooperation between Microsoft's Xbox and the games developer Bungie Studios, and reached Halo 3 in the last quarter of 2007. To give some indication of the popularity of this series, Halo 2, launched in 2004, has sold more than seven million copies worldwide so far. In all, 14.7 million copies of Halo titles have so far been sold, and more than 800 million hours of the online element of the Halo 2 game have been played.

Halo signifies the construction of world on world. It is a series of terraforms, models of possible worlds. This seems extraordinarily important to me in that it presages the kind of world that is now coming into being, one based on new disciplines like reflexive architecture, interaction design, environment art, and various forms of gaming that aim to redesign interaction. These disciplines allow passions that would have been difficult to express collectively to come into being through the design of new kinds of environments—synthetic worlds—that both facilitate play and close it down. I think it is no coincidence that there is currently so much attention being paid to new, more active forms of materiality: in a sense, these are the building projects of the twenty-first century because they presage a time when "there really is no barrier to a complete translation of every human interpersonal phenomenon on Earth into the digital space" (Castronova 2006, 48) with all manner of results, from new zones of economic activity through to new forums for interaction. After all, as Castronova (2006, 69) puts it, "what happens in these worlds is not just play, and not just communication. It is a complex thing, a combination of real interaction and a play-like context."

Thus, in Halo, the purpose of the game is to move the characters through vast outdoor and indoor environments that have been imagined in great detail. Though the environments are designed by concept artists and executed by teams of designers who want to make these worlds "look and feel real" (Trautmann 2004, 71), they are also open to fan feedback. The environments are themselves characters in the game, what the designers call "silent cartographers." Objects are always also locations. What we see is the construction of new fields of occurrences and the construction of new entities that can count as events (Newman and Simons 2007).

I want to suggest that Halo stands for a particular aspect of the modern world, namely, a shift in the nature of mediation toward "worlding" enabled by new material cultures that allow the affective priming of space to be systematized in ways that were not possible before. The game is symptomatic of the new "stickiness" that is now possible in three ways.

At one level, it stands for how modern business has moved on from a focus on producing objects to a focus on producing worlds that must also inevitably be spaces. Thus the business enterprise does not create its object but rather the world within which the object exists. As a corollary, the business enterprise does not create its subjects (as happened in the older disciplinary regimens) but rather the world within which the subject exists. As Lazzarato (2006, 188) puts it,

> the company produces a world. In its logic, the service or the product, just as the consumer or the worker, must correspond to this world; and this world in its turn has to be inscribed in the souls and bodies of consumers and workers. This inscription takes place through techniques that are no longer exclusively disciplinary. Within contemporary capitalism the company does not exist outside the producers or consumers who express it. Its world, its objectivity, its reality, merges with the relationships enterprises, workers and consumers have with each other.

The corporate aim is to produce and harvest what might be called decisive moments of affectively inspired semiosis, which can be played into through the redesign of environments. Such engrossing moments have a deeply engrained cultural history, of course. The decisive moment was, in large part, an invention of Renaissance painters trying to depict major turning points in history. They would build up scenes in great detail in which the disposition of every person and object counted as a part of a moment straining toward realization. The motif was subsequently taken up by photographers, and especially photojournalists. Famously, for Henri Cartier-Bresson, the decisive moment (the title of his exhibit at the Louvre, the first photographic show ever to be so honored) is the instant when a shutter click can suspend an everyday event within the eye and heart of the beholder, producing a confluence of observer and observed. It is the "simultaneous recognition, in a fraction of a second, of the significance of an event as well as the precise organization of forms which gives that event its proper expression" (Cartier-Bresson, 1952). Then the decisive moment is still very much a mainstay of modern drama and, most obviously, film. Cinema can be understood as a series of practical meditations on summoning up decisive moments: "truth 24 times a second," as Godard (cited in Mulvey 2006, 15) once put it. Cinema is able to produce not just speed but delay and deferral, preserving the moment at which the image is first registered in a kind of extended present.

On another closely related level, I argue that this game is symptomatic of the general rise of suggestible environments that can act to concentrate and guide infectious encounter by constructing traps for the affective flow of everyday life.[21] In turn, encounter can become specie, an insight that is drawn from the final writings of Althusser (2006), in which he refers to the genesis of a state of encounter, in which encounter is more and more able to be engineered so that it can be thought of as a kind of currency with a face value. But perhaps, rather than drawing on a monetary metaphor, a metaphor of cultivation might be more appropriate. For there seems to me to be a direct line of descent between the knowledges of semiotic arrangement and disposition that landscape gardeners like Humphrey Repton thought to be so crucial to their art of making fictions manifest and the games of today. These knowledges of arrangement and disposition are currently going through a new round of both strengthening and extension, as evidenced by, for example, the general rise in cartographic awareness in all spheres of life and most especially by the experimentation with new forms of interrelation between mapping and the senses (e.g., Jones 2006), which are allowing infectious relationships to be both represented and engineered as never before.

On a final level, these heavily corporatized suggestible environments signify a new sense of narrative that is not linearized (Fleming 2001). A good example of this sense of narrative is provided by many modern game-influenced movies. Take *Pirates of the Caribbean: Dead Men's Chest:*

> The film has no concern with cogent storytelling, and neither do today's youngsters. For them, fiction, like gaming, is an eternal present and plots a perpetuum mobile. The only narrative is to get to the next level. So while Pirates 2 spools for older people like a story whose reels have been muddled—a nightmare of botched narrative—for children and young adults, up to say, 20, the film advances to higher things on stepping stones of incremental surrealism. (Andrews 2006, 51)

Derrida spent a considerable period of his career considering the way in which writing had imposed a particular form of linearization of time and space on the world, which was, in effect, the infolding of space and time known as "book." But Derrida was at pains to point out that linearization represents "only a particular model, whatever might be its privilege," and he notes the increasingly obvious inadequacy of this model of arrangement to the "delinearized temporality" and "pluri-dimensionality" of contemporary thought: "what is thought today cannot be written according to the line and the book, except by imitating the operation implicit in teaching modern mathematics with an abacus" (Derrida 1998, 87). The linearization provided by writing and its consequent inadequacy for certain kinds of thought can be thrown into relief by writing schemes that do not deploy the linear norm, so-called nondiscursive writings. There are many of these emblematic genres. For example, take the language of flowers, an early modern schema that used a variety of somatic registers—layout (e.g., the circumference), color, texture, smell—to display the special indistinction between natural objects and rhetorical figures (Fleming 2001). This language made its way into many aspects of life as actual material objects, each of them understandable as utterances, from posy rings to nosegays, in a society that associated flowers with moral and other qualities. Viewed from our current perspective, such schemas as the language of flowers may appear to be inefficient codes, impoverished by a general lack of grammar and an unregulated three-dimensional, multisensory syntax that "cannot be further combined into a restricted and therefore consequential utterance" (Fleming 2001, 21).

But equally, from the perspective on linear narrative that is offered by some of the current developments, it is conceivable that a new form of narrative will be generated that is very close in form to the premodern prototype, one in which *conviction is carried in material objects and actions* (rather than what today is called the mind). Furthermore, this is a sense of the world in which nondiscursive writing is not readily distinguished from other human-made or naturally recurring patterns, wherein lies the recalcitrance to full referentiality that constitutes its particular force.[22]

This ambition is incarnated in Erasmus's celebrated description of a country house in which the walls, doors, galleries, flower beds, and wine cups are all decorated with improving messages, an imaginative development intended to move on from the extant holders and transmitters of religious knowledge like the stained glass window to something all-encompassing:

> "Who could be bored in this house," asks one guest, when among so many painted forms there is "nothing inactive, nothing that is not saying or doing something?" Writing is positioned throughout the house and gardens to catch at the eye and activate the memory: religious texts and images remind the host and his guests of the way to salvation, and encourage them to pray; emblematic plants and animals carry various moral lessons; and painted birds and other *trompe l'oeil* effects cause wonder at "the cleverness of nature . . . the

> inventiveness of the painter, [and] in each the goodness of God." (Fleming 2001, 139–40)

And it has never quite left the world. As Derrida (1998, 128) put it, the linear norm "was never able to impose itself absolutely," not just because acts of cognition can occur outside it but because the linear norm is set to function as a limit and so opens the very questions it appears to close: the contingencies of graphic phoneticism, and the philosophical system that relies on it, depend on an imposition that leaves in its wake all kinds of out-of-sequence gatherings that cannot be made to fit and that might be made to remind readers of the material practices that went into the production of the text. The ambition was kept in gardening, in some aspects of folk design, in parts of architecture. But there is more to this fugitive history. To illustrate this, I want to begin with Charleston House in Sussex, the famous home of the Bloomsbury set, notable especially for Vanessa Bell and Duncan Grant's rich decorative style. Inspired by Italian fresco painting and the Post-Impressionists, the two artists decorated the walls, doors, furniture, and garden at Charleston to the extent that the house and garden became a living work of art in which every surface was semiotically enhanced, thus reproducing Erasmus's dream of a country house that would speak out from every corner. In doing so, Bell and Grant produced a mock-up of what the modern world is becoming like, a space in which even the marginalia are semiotically charged.

But whereas their house and garden was an imaginative booster rocket that could be regarded as largely positive within its own bounds, much the same kind of ambition can also have profoundly negative consequences, as many totalitarian states have proved since.[23] On this dark side, what is crucial to understand is the degree to which so much of the modern world consists of marginalia made central by so-called reactionary modernist forces.

Of course, since the earlier part of the twentieth century, new visual technologies have run riot, technologies that extend both the means of representation (as in the proliferation of screens, the wall newspapers of the twenty-first century) and the registers that it is possible to decorate with images (as in the inhabitation of the precognitive domain by sigils like brands), thereby producing something closer to an electronic version of Erasmus's house in which every surface gives off continuously modulated messages such that an exchange of qualities, rather than just a transmission of information, takes place—what Bruno (2004, 7, 12) suggestively calls a "pandemic of images" that produces an "aggregate mnemonic structure" that consists of multiple levels and planes of stimulation, disposition, and recollection, all jumbled together in various living reappropriations that constitute a kind of choreography, rising and falling to rhythms of its own.[24]

There is evidence to suggest that this process is gathering pace as a result of the intervention of large-scale parallel and distributed computation in all its forms (Rotman 2000), which has allowed previously separate visual media—live action cinematography, graphics, still photography, animation, three-dimensional computer animation, typography, and so on—to be *combined* in novel ways, producing what Rotman (2000) calls "rampant visualism" and Manovich (2006) calls the "velvet revolution." The underlying logic of this revolution, which produces new media forms out of combination, is one of remixability in which the computer simulates all media, thereby inducing a transformation of visual language toward "motion graphics," that is, "designed non-narrative, non-figurative based visuals that change over time" (Frantz, as cited in Manovich 2006, 8). What counts is the arrangement of elements, such as size, aspect, a line of type, an arbitrary geometric or other kind of form, and so on, into a kind of dance: "we can compare the designer to a choreographer who creates a dance by 'animating' the bodies of dancers—specifying their entry and exit points, trajectories through space of the stage, and the movements of their bodies" (Manovich 2006, 12).[25]

This is, as I have tried to make clear, much more than some putative society of the spectacle, that is, an intensified deployment of the apparatus of the production of appearances. Rather than such a state of fallen grace, what I am trying to describe here is a reinhabitation, one based on making the environment—a word that itself becomes a contested one under the new conditions—into a semiotic soup, but one in which most of the signs are nondiscursive. This reinhabitation is akin to the biosemioticians' notion that the basic unit of life is the sign (Wheeler 2006).[26] It is not a direct imposition on a passive substrate of humanity but rather a reworking of what counts as through, a processual "haptic spatiality" (Hansen 2004). In other words, a nonrepresentational mode of writing that utilizes nonlinear deployments of time and space is again gaining a place in the world, with clear effects on what we regard as perception.

In turn, this change provides a pressing political task in a society in which this rampant visualism is coming into being. It poses obvious risks—but it also provides opportunities for building new kinds of locative community.

The Third Halo: Constructing New Affective Affinities

In its third manifestation, the halo is a standard scientific term, used to denote various optical phenomena that appear around light sources. It is therefore a natural term in sciences like meteorology, physics, and astronomy. For example, the galactic halo is a region of space surrounding spiral galaxies, including our galaxy, the Milky Way. It consists largely of old stars, gas, and dark matter. It is believed that the galactic halo is largely a consequence of the evolution of subgalactic clumps seeded from cold dark-matter density fluctuations.

I want to use this image to return to the *matter* of imagination, understanding materiality as a series of occasions that are always moments of knowledge. At the same time, I also want to think about the changing shape of knowledge itself after the onset of information technology, as knowledge increasingly takes on a significant nonparadigmatic halo that cannot be centered[27] or made a part of the whole, the result of not just the expansion of knowledge but its increasing ownership by communities that have little or no relation to formal knowledge structures. In other words, in this chapter, the third manifestation of the halo is as a whole series of knowledges thought to be of little or no consequence that form clumps of various kinds, a background that turns out to be fundamental in seeding the universe of knowledge rather than incidental. Why? Because I am convinced that these petty knowledges are a resource that can be tapped to form a new political genre or genres,[28] one that calls to and relies on affective contagion and that might be used to reengineer affective qualities like bravery and courage in productive ways. This politics consists of clumps of like minds arranged loosely and indistinctly in semidirected practices that move beyond understanding affective contagion as simple contact toward understanding affective contagion as a kind of fluency practiced by design. Let me make it clear: this is not to suggest that if these practices were aggregated, they would suddenly form a new political force, but rather that they can form a different kind of choreographic strain, a contrary motion that both works with and against the grain of "being-toward-movement" and that might allow us to sense and even construct new affective strains.

Set against those who think that "our stunted imaginations have largely lost the ability to think what a society other than capitalism . . . might look like" (Smith 2007, 2), I think we are living in a time of extraordinary imaginary outbursts if only we have the nous to touch and feel them, imaginary outbursts founded in the cooperative symbiosis provoked by new situations, imaginary outbursts that force thinking by producing *affective affinities*. These outbursts have already had considerable purchase on the world of mass daily practices, but on the whole, we are not picking them up because they are based on "discontinuities of pattern, the tiny causalities of chance, the reparative and tender (as opposed to deadly and terrifying) features of intricate connection" (Orr 2006, 18). They do not fit the standard categories—active/passive, micro/macro, passion/calculation, interested/disinterested, objective/belief—we use to describe the world.

These outbursts could be named in all kinds of ways, no doubt. But I want to draw on modern performance studies to try to describe them in more detail. Performance has always understood the power of affective contagion and sometimes has highlighted it. Think only of Artaud's alchemical theater:

> What modern social science tried to make intelligible, Artaud tried to make real: the contagion of gesture, the communicative power of a scream, a mimetic theatre of collective seizure and frenzied emotion, Artaud's intent was not to start a panic but experiment through performance with features of the social—never far from the alchemy of the theatre—that collective terror also opens toward.

> "The mind's capacity for suggestion" which Artaud identifies as one source of theatre's transformative power, is precisely the capacity that modern social science locates as one source of the social itself. (Orr 2006, 8)

I will call these outbursts "dances that describe themselves" (Foster 2002) to give me a means of naming them and as a place to start from, as a piece crafted spontaneously in performance—in the moment. The phrase comes from the work of that well-known dancer-choreographer of improvisation, Richard Bull, and his allegiance to thinking on your feet by choreographing while you dance, thus producing a leaderless community. But his Dance was not just a piece of random improvisation, worshipping "liveness." Far from it. It was an act of possession—and command. The premise was that a set of dancers would come together and over several weeks slowly tune their worldview to the presence of the Dance That Describes Itself. This tuning was intended to unite their bodies (and, to an extent, those of the audience)[29] in the flexing, undulating mass of the Dance:

> They moved in and out of "possession," enjoying the shift in perspective, the different sense of agency that becoming inhabited by the dance allowed. The Dance told them what to do and they necessarily complied, yet they also created The Dance, determining when and how it might take control of their dancing. The contradiction between these two selves, one possessed and one in command, opened an ironic tension that reverberated throughout the entire performance, a tension compounded by the fact that the dancers described, often with clinical precision, their actions.
>
> Typically, the act of possession entails a loss of speech and the inability to describe during or afterwards what happened while dancing. *The Dance That Describes Itself* plays upon this venerable and ancient trope of giving oneself to the dance, becoming one with the dance, or being free in the dance. Yet it constructs a different kind of possession. Dancers are asked to remain highly conscious of their circumstances and to describe their actions verbally. Rather than serve as mute embodiment for cosmic forces, indescribable

> in their proportion and power, these dancers comment adroitly on mundane motives, frustrations, or desires. The collision between two incommensurate images of dance—one speechless and transcendent, the other analytic and pedestrian—reinforces the irony inherent in the initial proposition of being possessed. (Foster 2002, 11)

I would argue that the practices that I want to describe are akin to this stance in that they involve a careful tending of knowingness through the design of empowering situations based on producing new and speculative *locatives*, indications of place *and* direction *and* affinity, which privilege an openness of form that is still, however, able to shape and mold and comment on that process. At the same time, these locatives produce new time frames, new notions of "calendarity."

What would these new locatives look like? The history of performance undoubtedly gives us some clues, where performance is understood as the construction of socially and technically informed living entities, since in many ways it has been born out of an impulse to remap spaces and, in particular, to escape the constraints of enclosed theatrical spaces and the kind of conventions they abide by: discrete physical locations exploiting particular kinds of sound and lighting, linear manipulation of timelines through devices like reminiscence and premonition. Indeed, Norman (2006) argues that the vestigial geometries of these spaces and times still hamper our ability to craft other kinds of social encounter. We keep on beating the same bounds, often unconsciously.

Perhaps the most important step has been to get away from an obsession with exact localization. In the early nineteenth century, in the famous preface to *Cromwell*, Victor Hugo (1827) had already pointed out the benefits of the strategy of localization in inducing a sense of reality, but also its risks in dictating imaginative content. The problem, of course, is that too often, the reaction to the risks associated with localization has been to fall back on a notion of spontaneous gatherings of individuals, along with the common graphic vernaculars for depicting these instant multitudes—the crowds of which various forms of the flash mob are often considered to be the latest manifestation (Schnapp and Tiews 2006). The generation of apparently primal spontaneity has had a long history in performance, dating from at least Wagner's "free associations of the future," and it has had obvious political downsides. However, at the same time, it has also led to a very large amount of thought about how performance works at the preindividual level and how the performer acts as an enhanced transmitter of various forms of sympathy, culminating in many acts of ecological theater that try to conjure up a sentient unconscious, if that is not a contradiction in terms, through creative engagement of the feelings of the audience in the exploration of space (Banes and Lepecki 2006).

The tension between these strategies of localization and spontaneity, and the knowledges that they produce, is currently being worked out in "postdramatic" (Lehmann 2006) artistic performances that pull all sorts of beings into a communion of direct living perception that develops with and within time (Zielinski 2005). In particular, these performances explore the dynamics of affective emergence by constructing "living organizations" out of the new locative media. That task involves maximum experimentation across many registers of the senses (Banes and Lepecki 2006) to "feel" all the data available in a particular universe that might belong to an emerging entity,[30] using the full range of modern locative technologies as vital intermediaries. In turn, this task has generated what is often a calculated indifference to where performance is meant to take place. For example, performance can even be located in outer space, in the domain of so-called metagestural proxemics, as in the space suit that will be crammed with communication electronics and thrown out of the International Space Station to burn up in the atmosphere, or tometaxy.net's attempt to produce a collective public sculpture to world peace in orbit around the Earth and ultimately a moon installation or Nam June Paik's moon. In other words,

> locative media performances encompass participants and forge identities ranging from the most intimate to the most distant; the propensity of such performances to go global is equalled by their aptitude to inject highly localized, often time-bound events into overall connected fabrics. It is this tension between localized input on the one hand and web-borne, purportedly universal resonance

on the other, that gives mobile systems their complex social and artistic potential. (Norman 2006, 4–5)

In practice, this means that postdramatic theater attempts to produce performances across many sites simultaneously in what we might call a choreopolitics that is backed up by a range of different practices that blur the divides between what used to be known as art, as politics, as social science method, and as information technology, in a concerted attempt to batter down particular imaginative resistances through a mixture of "displacement, dislocation, distribution, and disorientation" (Jones 2006, 3).[31] This blurring first occurred in the 1960s, but after a forty-year history, it now provides a body of formal and informal knowledge of considerable sophistication, ranging from the kind of project that simply utilizes user-led functional cartography (e.g., Michel Teran's *Life: A User's Manual,* which tries to technologize Perec) through Jonah Brucker-Cohen and Katherine Moriwaki's UMBRELLA.net, and on to all manner of projects that chart out movement (such as Teri Rueb's *Choreography of Everyday Movement,* where trails of dancers moving through the city are tracked with GPS to obtain real-time dynamic drawings that are then printed onto acetate sandwiched between stacked glass plates that grow taller and more complex with each addition). One might argue that such projects act as nothing more than twinkling marginalia. That would, I think, be a mistaken judgment; rather, I think they are attempts to call up the outlines of spaces that can produce affective shifts which, though they may appear to be minor, can have major effects, for example, spaces that administer a shock to irritation so that it moves from a space in which overannoyance is accompanied by insufficient anger to something more affectively productive—such as the bravery required to actually intervene (Ngai 2005; Gallop 2006).

To end, I will fix on just two examples of this new form of mobile acting up, examples that tap into a long tradition of artists acting as agents provocateurs investing wider communities (Hampton 2007), examples that aim to produce politically charged works of art that exist outside any extant paradigm as a kind of kinetic outcropping or even stray that exhibits no deeper truth than the achievement of an affective phase shift (Barthes 2005). But that, of course, is the point.

The first example is the Displaced Emperors project. Displaced Emperors was the second relational architecture project. This installation used an "architact" interface to transform the Habsburg Castle in Linz, Austria. Wireless three-dimensional sensors calculated where participants pointed to on the facade, and a large, animated projection of a hand was shown at that location. As people on the street "caressed" the building, they could reveal the interiors, which corresponded to Chapultepec Castle, the Habsburg residence in Mexico City. In addition, for ten schillings, people could press the "Moctezuma button" and trigger a temporary postcolonial override consisting of a huge image of the Aztec headdress that is kept at the ethnological museum in Vienna.

The second example is Ballettikka Internettikka. Ballettikka Internettikka is a series of tactical art projects that began in 2001 with the exploration of Internet ballet. It explores wireless Internet ballet performances combined with guerrilla tactics and mobile live Internet broadcasting strategies. Its Internet guerrilla performance has mainly consisted of invasions of particular art houses, such as the Bolshoi Theatre in Moscow (March 2002), La Scala in Milan (November 2004), and the National Theatre in Belgrade (October 2005), and the subsequent mounting of alternative Internet performances within their confines, on the premise that the passions of opera can be transferred to the "passionless" Internet, so producing radical emotions. In 2006, the two main artists, Stromajer and Zorman, performed a new guerrilla net ballet, this time in the men's toilet in the basement of Volksbühne Berlin, using a flying cow, a group of small robots—and a toilet seat. The artists utilized low-tech mobile and wireless equipment for the invasion and live broadcast: a public, unprotected wireless Internet connection point, available for free at the Rosa-Luxemburg Square in Berlin, and free RealProducer and Live LE software for streaming video and audio live manipulation.

In one sense, interventions like these are simply the latest chapter in the long history of trying to produce grammars of movement of the kind to be found in the history of dance since at least the sixteenth century. But at another level, they are embryonic political interventions, affective utterances that, through the production of spatial and temporal coherences that are also new forms of imaginative assay,[32] are intended to boost encounter and thereby provide new means of animatedness and attentiveness. In particular, they are trying to struggle out from under a notion of a place that is bounded—an "environment"—toward notions of place as relationships with space that are rather like the face in their ability to be expressive and to reveal what the other is thinking: space as "the eyes of the skin," so to speak (Pallasma 2005). Thus space becomes richly emblazoned with signs of thought. In a sense, space becomes face.

Notes

I would like to thank Søren Buhl, Armin Geertz, Susan Hurley, Britta Timm Knudsen, Sally Jane Norman, and Minna Tarka for their comments on this chapter.

1 Of course, the imagination has routinely been figured as something that cannot be conventionally controlled, but there are many reasons to think that this is only partially the case and that it can indeed be engineered, especially in an age of manufactured vision in which "we now know ourselves in our mind's eye mostly by projecting a camera's eye view" (Warner 2006, 6).

2 Imagination actually fires in the same area of the brain as belief and can generate equally strong affective reactions. Thus most philosophers now count it as a "distinct cognitive attitude," different from beliefs in some respects but not others.

3 I take representation here to be able to be understood as doings rather than a relation between an inner and an outer. Our basic grip on the world consists not of inside out but of representing deeds: deeds are themselves representational (Rowlands 2006).

4 And achievement of the kind that I want to describe requires diplomacy. As Stengers (chapter 1) puts it, intent on describing the practice of shuttling between parties that disagree, "the art of diplomacy does not refer to goodwill, togetherness, the sharing of a common language, or an intersubjective understanding. Neither is it a matter of negotiation between flexible humans who should be ready to adapt as the situation changes. It is an art of artificial

arrangements that do not exhibit a deeper truth than their very achievement—the event of an articulation between protagonists constrained by diverging attachments and obligations in situations where contradiction seems to rule, a rhizomatic event without a ground to justify it, or an ideal from which to deduce it."

5 As the Roman emperors began to think of themselves as divine beings, they wore a crown in public to imitate the sphere of light from the sun.

6 Here the halo could have a very detailed iconography. Thus round halos were used to signify saints. A cross within a halo was used to signify Jesus. Triangular halos were used for representations of the Trinity. Square halos were used to depict unusually saintly living persons, still bound to earth and so not able to obtain the perfection of the circle. Allegorical figures such as the virtues wore hexagonal haloes (Fisher 1995).

7 But it is important to point out that the halo is not confined to Western religious history. For example, it has been widely used in various forms of Buddhist iconography since at least the first century A.D. to depict the Buddha or Buddhist saints, a direct importation from the West to the East.

8 The example is taken from Sloterdijk (1998).

9 As this instance shows, the idea that a politics of radical difference has to entail a choice between networks of signification or networks of embodied matter seems overdone; rather, recent work argues that embodied matter always has sign content (Wheeler 2006).

10 It is now generally accepted that the brain developed in response to and as a function of social interaction and especially the ascription of intention—that is, the attribution of actions, motives, intentions, and beliefs to fellow interactants—and that what we perceive is set up by the wiring of interaction produced by the set of most notably human abilities that plausibly evolved together, all of which were boosted by the enhanced communicative interaction arising from the paraphernalia of language: parsing, turn-taking, repair, and the like. The brain, in other words, has become an instrument of shared activity—an interaction engine (Enfield and Levinson 2006)—rather than an individual setting. And within broad parameters, this shared activity is remarkably heterogeneous, aided by the fact that the brain is in any case plastic so that particular experiences of shared activity act in particular ways, by the fact that systems of shared activity

generate emergent properties and by the fact that cultural variation is therefore more than just incidental but central to interaction.

11 It is important to note that in this chapter I will be taking embodiment to be a linked, hybrid field of flesh and accompanying objects rather than a series of individual bodies, intersubjectively linked. I take the presence of objects to be particularly important because they provide new means of linkage (Zielinski 2005)—new folds, if you like.

12 It may even be, following Tarde, that memory and habit are forms of imitation: "engaged in either, we in fact imitate ourselves, instead of another person: memory recalls a mental image, much as habit repeats an action" (Potolsky 2006, 116).

13 This distinction can be traced back to Aristotle.

14 Any Web search for courage or bravery immediately produces vast numbers of military examples, showing the centrality of this conception of courage and bravery to our judgments.

15 Aristotle does not mean that bravery is simply an average; rather, the golden mean is different for each person, depending on her character and situation.

16 I could no doubt have chosen other examples—the warlike honor code of the Pushtun comes to mind, as does the history of Gandhian nonviolence, but these two examples seem to me to be striking enough to make the point.

17 Counting coup could also mean taking an enemy's weapons while he was still alive; striking the first enemy to fall in battle, no matter who killed him; stealing a horse tied up in an enemy's camp; and so on.

18 I am thinking here of movements like the International Solidarity Movement, which produced "passive" heroes and heroines like Tom Hurndall and Rachel Corrie. "Their acts of solidarity articulate a practical riposte to the despairing twentieth-century voices that wanted to discredit this sort of gesture by arguing that the openness and undifferentiated love from which it derives is tainted, ignoble, and unpolitical" (Gilroy 2004, 90). It also points to another halo: the HALO Landmines Trust.

19 Halo was also a fictional superheroine published by DC Comics in the 1980s and 1990s.

20 A film is also planned for 2008, with the involvement of Peter Jackson.

21 To some degree, large cities have always acted in this way, allowing "epidemics" of imitation to be marshaled and directed, but what I am suggesting is something with a much greater element of design.

22 This is, of course, a sense of the world that has long been familiar to anthropologists and archaeologists in cultures where symbolism and daily life are intertwined in a network of entanglements that are both means of empowerment and dependencies, typified by Hodder's (2006) recent study of Catalhöyük with its sheer amount of elaborate wall art, stimulated by a lime-rich plaster that needed continual resurfacing and might be thought of as a prototype of the constantly refreshed screen.

23 Thus Herf (2006, 274) notes the way in which Nazi Germany attempted to design environments that would produce a total political experience by using media like radio, mass meetings, print, and especially weekly wall newspapers that "stared out at the German public for a week at a time in tens of thousands of places German pedestrians were likely to pass in the course of the day."

24 As I have tried to point out, the difference with the past is that these images are able to be interactive: their calls can produce responses that can act on their nature.

25 Such developments arise out of new practices combining with new outlets of expression (e.g., most recently, YouTube).

26 Indeed, I think that it is quite possible to historicize the argument of Hoffmeyer and others that "signs, not molecules, are the basic units in the study of life" (cited in Wheeler 2006, 123), making it into a symptom of the present.

27 Indeed, one might argue that there is no longer a center, only a halo.

28 E.g., it would be possible to take a leaf from the art of landscape gardening again and suggest that what has become crucial is a knowledge of arrangement or disposition, of finding (search), which is currently going through a convulsion as the kinds of nonlinear, nondiscursive thinking and representation that I outlined in the previous section are brought back into play. Gardening may seem an odd metaphor from which to work, but I am sure that it fits the bill—passionate, sensuous, self-evolving, multisensory, synaesthetic, the favorite of Klee (Harrison, Pile, and Thrift 2004; Tilley 2006), the premier art of *cultivation*.

29 "As dancers open their physicalized imaginations to entertain the possibility of any and all next actions, they also track the results of acting upon or rejecting those impulses. As viewers watch, going with the flow of events, they also critically engage with that going.

Throughout the performance, both dancers and viewers ask themselves, what is going to happen next? And what difference will it make to this performance's significance?" (Foster 2002, 16).

30 The link to Whitehead's notion of prehension is clear.

31 E.g., think of the ubiquity of Chernoff faces, which use facial features to represent data, or the growth of "choreogenetics" (Lapointe 2005), which uses genetic algorithms to generate choreographic sequences.

32 On another level, these performances are a part of a general tendency to "move the mutual implication of actors and spectators in the theatrical production of images into the centre" (Lehmann 2006, 186).

References

Abrams, Janet, and Peter Hall, eds. 2006. *Else/Where: New Cartographies of Networks and Territories*. Minneapolis: University of Minnesota Press.

Althusser, Louis. 2006. *Philosophy of the Encounter: Later Writings, 1978–1987*. London: Verso.

Andrews, Nigel. 2006. "The Guess Men." *FT Magazine*, November 11.

Banes, Sally, and Andre Lepecki, eds. 2006. *The Senses in Performance*. New York: Routledge.

Barthes, Roland. 2005. *The Neutral*. New York: Columbia University Press.

Bermúdez, José. 2003. *Thinking without Words*. Oxford: Oxford University Press.

Bickmore, Timothy W., and Rosalind W. Picard. 2005. "Establishing and Maintaining Long-Term Human–Computer Relationships." *ACM Transactions on Computer–Human Interaction* 12: 293–327.

Blum, Deborah. 2007. *Ghost Hunters: William James and the Search for Scientific Proof of Life after Death*. London: Century.

Brennan, Teresa. 2004. *The Transmission of Affect*. Ithaca, N.Y.: Cornell University Press.

Bruno, Giuliana. 2004. "Modernist Ruins, Filmic Archaeologies." In *Free and Anonymous Monument*, ed. J. Wilson and L. A. Wilson, 1–7. Newcastle upon Tyne, U.K.: Baltic.

Cartier-Bresson, Henri. 1952. *The Decisive Moment*. New York: Simon and Schuster.

Castoriadis, Cornelius. 1997. *The Imaginary Institution of Society*. Cambridge: Polity Press.

Castronova, Edward W. 2006. *Synthetic Worlds: The Business and Culture of Online Games*. Chicago: University of Chicago Press.

Collectivo Situationes. 2005. "Something More on Research Militancy: Footnotes on Procedures and in Decisions." *Ephemera* 5: 602–14.

Deetz, James. 1996. *In Small Things Forgotten*. New York: Anchor.

Derrida, Jacques. 1998. *Of Grammatology*. Trans. Gayatri Spivak. Baltimore: Johns Hopkins University Press.

Eilan, Naomi, Christoph Hoerl, Teresa McCormack, and Johannes Roessler, eds. 2005. *Joint Attention: Communication and Other Minds*. Oxford: Oxford University Press.

Eliasoph, Nina. 1998. *Avoiding Politics: How Americans Produce Apathy in Everyday Life*. Cambridge: Cambridge University Press.

Enfield, Nicholas J., and Stephen C. Levinson, eds. 2006. *Roots of Human Sociality: Culture, Cognition, and Interaction*. Oxford: Berg.

Fisher, Sally. 1995. *The Square Halo and Other Mysteries of Western Art*. New York: Harry N. Abrams.

Fleming, Juliet. 2001. *Graffiti and the Writing Arts of Early Modern England*. London: Reaktion.

Foster, Susan L. 2002. *Dances That Describe Themselves: The Improvised Choreography of Richard Bull*. Middletown, Conn.: Wesleyan University Press.

Gallop, Jane, ed. 2006. "Envy." Special issue of *Women's Studies Quarterly* 34, nos. 3–4.

Gell, Alfred. 1998. *Art and Agency: An Anthropological Theory*. Oxford: Oxford University Press.

Gilroy, Paul. 2004. *After Empire: Melancholia or Convivial Culture?* London: Routledge.

Goldman, Alvin I. 2006. *Simulating Minds: The Philosophy, Psychology, and Neuroscience of Mindreading*. Oxford: Oxford University Press.

Gordon, Scott P. 2002. *The Power of the Passive Self in English Literature, 1640–1770*. Cambridge: Cambridge University Press.

Grandin, Temple. 2005. *Animals in Translation*. London: Bloomsbury.

Gross, Daniel. 2006. *The Secret History of Emotion: From Aristotle's Rhetoric to Modern Brain Science*. Chicago: University of Chicago Press.

Gumbrecht, Hans. 2004. *Production of Presence: What Meaning Cannot Convey*. Stanford, Calif.: Stanford University Press.

Hampton, Howard. 2007. *Born in Flames: Termite Dreams, Dialectical Fairy Tales, and Pop Apocalypses*. Cambridge: Harvard University Press.

Hansen, Mark. 2004. *New Philosophy for New Media*. Cambridge: MIT Press.

Harrison, Stephan, Steve Pile, and Nigel Thrift, eds. 2004. *Patterned Ground*. London: Reaktion.

Henare, Amiria, Martin Holbraad, and Sari Wastell, eds. 2007. *Thinking through Things: Theorising Artefacts Ethnographically*. London: Routledge.

Herf, Jeffrey. 2006. *The Jewish Enemy: Nazi Propaganda during World War II and the Holocaust*. Cambridge: Harvard University Press.

Hodder, Ian. 2006. *Catalhoyuk: The Leopard's Tale*. London: Thames & Hudson.

Hugo, Victor. 1827. Preface to *Cromwell*. New York: French and European, 1989.

Hume, David. 1739. *A Treatise of Human Nature*. Repr., Harmondsworth, U.K.: Penguin, 1985.

Hurley, Susan, and Matthew Nudds, eds. 2006. *Rational Animals?* Oxford: Oxford University Press.

Jones, Caroline A., ed. 2006. *Sensorium: Embodied Experience, Technology, and Contemporary Art*. Cambridge, Mass.: MIT Press.

Jones, Colin. 2000. "Pulling Teeth in Eighteenth-century Paris." *Past and Present* 166: 100–45.

Lapointe, François-Joseph. 2005. "Choreogenetics: The Generation of Genetic Mutations and Selection." *ACM GECCO* 5: 366–69.

Lazzarato, Maurizio. 2006. "The Machine." Transform.eipcp.net. http://transform.eipcp.net/transversal/1106/lazzarato/en.

Lear, Jonathan. 2006. *Radical Hope: Ethics in the Face of Cultural Devastation*. Cambridge: Harvard University Press.

Lehmann, Hans-Thies. 2006. *Postdramatic Theatre*. London: Routledge.

Manovich, Lev. 2006. "After Effects, or Velvet Revolution in Modern Culture," part 1. http://www.manovich.net/DOCS/motion_graphics_part1.doc.

McNeill, Daniel. 1998. *The Face: A Guided Tour*. London: Hamish Hamilton.

Mulvey, Laura. 2006. *Death 24× a Second: Stillness and the Moving Image*. London: Reaktion.

Newman, James, and Iain Simons. 2007. *100 Videogames*. London: British Film Institute.

Ngai, Sianne. 2005. *Ugly Feelings*. Cambridge, Mass.: Harvard University Press.

Nichols, Shaun, ed. 2006. *The Architecture of the Imagination: New Essays on Pretence, Possibility, and Fiction*. Oxford: Oxford University Press.

Norman, Sally J. 2006. "Locative Media and Instantiations of Theatrical Boundaries." *Leonardo* 14, http://www.leonardo.info/isast/journal/.

Orr, Jackie. 2006. *Panic Diaries: A Genealogy of Panic Disorder*. Durham, N.C.: Duke University Press.

Pallasma, Juhani. 2005. *The Eyes of the Skin*. Chichester, U.K.: John Wiley.

Parkes, Donald N., and Nigel J. Thrift. 1979. "Time Spacemakers and Entrainment." *Transactions of the Institute of British Geographers* NS4, no. 3: 353–72.

———. 1980. *Times, Spaces, Places*. Chichester, U.K.: John Wiley.

Pettit, T. 2003. "Ritual and Performance in Theatre." Unpublished paper.

Potolsky, Matthew. 2006. *Mimesis*. London: Routledge.

Rath, Richard. 2003. *How Early America Sounded*. Ithaca, N.Y.: Cornell University Press.

Rotman, Brian. 2000. "Going Parallel." *SubStance* 29: 56–79.

Rowlands, Mark. 2006. *Body Language: Representation in Action*. Cambridge, Mass.: MIT Press.

Schnapp, Jeffrey T., and Matthew Tiews, eds. 2006. *Crowds*. Stanford, Calif.: Stanford University Press.

Sloterdijk, Peter. 1998. *Sphären I: Mikrosphärologie*. Frankfurt, Germany: Suhrkamp.

Smith, Adam. 1759. *The Theory of Moral Sentiments*. Repr., Oxford: Oxford University Press, 1976.

Smith, Neil. 2007. "Another Revolution Is Possible: Foucault, Ethics, and Politics." *Environment and Planning D* 25: 191–93.

Thrift, Nigel J. 2005. "From Born to Made: Technology, Biology and Space." *Transactions of the Institute of British Geographers* NS30: 463–76.

———. 2006. "Re-inventing Invention: New Tendencies in Capitalist Commodification." *Economy and Society* 35: 279–306.

———. 2007. *Non-Representational Theory: Space, Politics, Affect*. London: Routledge.

Tilley, Christopher. 2006. "The Sensory Dimensions of Gardening." *Senses and Society* 1: 311–30.

Tønder, Lars, and Lasse Thomassen, eds. 2005. *Radical Democracy: Politics between Abundance and Lack*. Manchester, U.K.: Manchester University Press.

Trautmann, Eric S. 2004. *The Art of Halo: Creating a Virtual World*. New York: Ballantine Books.

Walzer, Michael. 1988. *Spheres of Justice*. Oxford: Blackwell.

Warner, Marina. 2006. *Phantasmagoria: Spirit Visions, Metaphors, and Media into the Twenty-first Century*. Oxford: Oxford University Press.

Wheeler, Wendy. 2006. *The Whole Creature: Complexity, Biosemiotics, and the Evolution of Culture*. London: Lawrence and Wishart.

Zielinski, Siegfried. 2005. *Deep Time of the Media: Toward an Archaeology of Hearing and Seeing by Technical Means*. Cambridge, Mass.: MIT Press.

PART III

Political Technologies: Public (Dis)Orderings

7 Front-staging Nonhumans: Publicity as a Constraint on the Political Activity of Things

NOORTJE MARRES

OVER THE LAST YEARS, a sizeable publicity machine has been set up by governments, energy companies, and environmental organizations to promote reductions in domestic energy consumption as a way for people to help "combat global warming."[1] These initiatives have been criticized on various grounds, not in the least because of the lack of credibility of their hyperbolic claims such as the assurance that fixing energy-efficient lightbulbs or routinely unplugging one's mobile telephone charger "helps repair the planet"[2]—claims that for a while were endlessly repeated on billboards, in the press, and so on, in the Netherlands and the United Kingdom. Perhaps most important, social critics have charged these media campaigns with trivializing the ideals of citizenship and public participation. Thus it has been pointed out that because of their focus on basic household interventions, as a way of making it "feasible" to do one's share for the climate, these environmental campaigns in effect redefine civic involvement as an atomized, isolated, and individualistic activity. They are then seen as "privatizing" citizenship to the point that effective intervention on the part of the public actually becomes less rather than more feasible (for a discussion, see Clarke et al. 2007). Interestingly, however, publicity campaigns seeking to "green" the home are equally vulnerable to almost the opposite criticism, namely, to the charge that they promote the invasion of private places by public authorities and thus amount to a "de-privatization" of the home. There is certainly no lack of concrete examples to support such a claim, such as the "DIY Repairs" communications initiative of the mayor of London, launched in June 2007, which offers free house visits by a "green homes concierge service" to provide practical advice on how to make your home more energy-efficient, and yes, to help "save the planet."[3] Around the same time, the department store M&S announced that its textiles will soon carry a new label: "Think Climate—Wash 30 C."[4] Considering the ubiquity of such attempts to insert environmental considerations into the fabric of everyday life, it certainly seems important to be able to draw on critical repertoires that allow us to question the intrusion of public authorities into intimate places. However, it seems equally important that such campaigns can be seen to *problematize* the understanding of citizenship and the distinction between the public and the private domain on which such critical repertoires tend to rely.

Projects that define the home as a site where people can do their bit for the climate can be said to challenge certain classic assumptions regarding the proper locations and formats for public involvement in politics. As Andrew Dobson and Derek Bell (2005) have pointed out, contemporary practices of environmental citizenship invite consideration of the special affordances of practices that are traditionally defined as private for engagement with public affairs. Thus they make it clear that one of the defining features of environmentalism is that the sphere of "the reproduction of everyday life" here comes to the fore as an important setting for citizenly action. For this reason, environmental practices can seem to scramble the neat geometry that provided the scaffolding for classic republican conceptions of citizenship, as in the work of Aristotle and Rousseau. The republican tradition firmly anchored civic action on one side of the divides between the public and the private domain, between matters of general concern and mere particularities, and between the lofty questions of the common good to which the leisurely classes dedicate themselves and the mundane troubles and worries that keep working men and women busy. These distinctions can easily start shifting around when considering environmental practices, and more specifically, the connections that climate change campaigns establish between this global issue and domestic energy practices. Moreover, such campaigns could be said to actively contribute to the production of confusion regarding the distinction between the public and the private realms. Thus it is possible to understand green-the-home campaigns like that of the mayor of London as an attempt to actively transform the intimate sphere of the household into a very public place indeed, and this not only in the sense that the home in these campaigns becomes subject to extensive attention from public entities like governments, news media, and their audiences. As mentioned, domestic practices here also come to be defined in terms of their impact on common goods like the global climate and the planet, and as private practices are thus evaluated in terms of their public effects, the former could be said to acquire a public aspect themselves.

It may obviously be necessary to take such unsettling effects of environmental practices on established political distinctions into account when seeking to evaluate these practices. In this respect, green-the-home campaigns draw attention to yet another feature of environmental practices that may also deserve consideration: in these campaigns, *material things* are placed in the forefront as crucial tools or props for the performance of public involvement in issues. Mobile phone chargers, thermostats, lightbulbs, and water cookers are here presented as so many "technologies of citizenship" (Rose 1999) that may equip individuals to practically intervene in, or at least relate to, global public affairs. These campaigns thus attribute special affordances to domestic technologies in terms of their ability to help bridge the divide between people "in here," in the home, and issues "out there." However, though it thus seems clear that the role of domestic technologies in the performance of environmental citizenship deserves appreciation, it is far from self-evident how we should conceptualize their role as mediators of public involvement in issues. The reasons for this should become clearer later, but it has to do with the exclusion of material things from civic practices in classic conceptions of citizenship alluded to earlier. According to the republican tradition, material practices clearly belong in the private, noncivic domain. Just as important, an instrumentalist explanation of the role of things in politics, which would straightforwardly define domestic technologies as neutral tools for problem solving, that is, for alleviating the causes of climate change, is not satisfactory for a number of reasons. Moreover, postinstrumentalist frameworks that have been developed in recent decades to account for the role of technology in politics are equally difficult to apply to this case. As I will discuss in this chapter, post-Foucauldian studies of the politics of technology have importantly drawn attention to the capacity of objects to mediate political relations, but in doing so they suggested that this capacity is predicated on their "clandestinity," that is to say, on the circumstance that technologies are *not* usually recognized as political agents. This requirement, however, clearly is not met in the case of publicity campaigns to green the home, in which domestic technologies feature as major protagonists on billboards, in the press, and so on. In this chapter, I therefore turn to the work of the pragmatist philosopher John Dewey to explore how to conceptualize the relation between publicity media, material practices, and public involvement in politics. In particular, Dewey's concept of the public provides a crucial conceptual resource for understanding how material things may acquire the capacity to mediate people's involvement in political affairs under conditions of publicity. Moreover, such a detour via the work of this classic pragmatist can also help to make clear how post-Foucauldian studies of technology contribute to the understanding of public involvement in politics.

Household Devices as Technologies of Citizenship?

Media campaigns that focus on doable interventions in the home can partly be understood as a particular solution to the *problem* that citizen involvement in climate change presents. Thus the literature on the public understanding of climate change has put much emphasis on the obstacles there are to the effective public communication of this issue, such as its scientific complexity and theoretical abstractness. These features have been widely understood as placing climate change at a great, perhaps unbridgeable distance from people's everyday concerns (Trumbo 1995; Weber 2006). Partly in response to this problem, several authors have pointed at the capacity of visual media to lift complex environmental affairs out of the domain of abstract scientific calculation and to transpose them into the realm of human experience (Allan, Adams, and Carter 1999; Jasanoff 2001). Resisting rationalist discourses that would exclude aesthetic and affective modes of concern from the "proper" registers of citizenship, these authors write affirmatively about the affordances of media like television, newspapers, and the Internet for the cultivation of environmental citizenship. Thus Szerszynski and Toogood (1999) have pointed at the opportunity that visual media provide for expanding the repertoire of civic concern with environmental problems to include sensory and emotive forms of sensibility, which are closer to lived experience. The focus on domestic practices as sites of environmental involvement can be understood in the light of further elaborations or radicalizations of such claims. Phil Macnaghten has argued that visual imagery of natural disasters, though appealing to the emotions, ultimately fail to inspire sustainable forms of environmental concern in people. Though such natural events may be within the realm of human experience, they too are distant from everyday life, and they do not create room for personal agency vis-à-vis environmental problems. Claiming that concern about environmental problems begins with *personal* experience, Macnaghten (2003, 80–81) has concluded that environmental publicity campaigns should "start from people's concern for themselves, their families and localities as points of connection for the 'wider' global environmental issues." Moreover, in his account, such an approach should involve a focus on feasible interventions: "people are seeking credible solutions, 'in bite sized chunks,' where the material effects of individual action become visible and enduring" (81). It seems no exaggeration to say that organizers of recent climate change campaigns that focus on the home have at least in some respects heeded this call. Thus the aim of the "DIY Repairs" campaign of the mayor of London "is to raise awareness of climate change in a positive 'can do' sense," and the organization has justified this orientation in reference to survey findings that Londoners are most likely to be willing to do something for the environment if this does not require much effort.[5]

A particular understanding of the challenge that environmental issues present for everyday people then seems to be involved in efforts to define the household as a site for the performance of "climate citizenship." However, the preceding sociological accounts of this challenge do not tell us very much about the role of material entities like lightbulbs and phone chargers in this regard. They situate citizenship somewhere between "phenomenology" and "agency," between the human experience of environmental problems and the practical opportunity to act on these problems. As Macnaghten suggests, by redefining environmental citizenship in terms of practical interventions in the lifeworld, citizenly action is displaced onto the plane of physical practice, where interventions have "material effects." However, in the case of climate change, the notion of material or physical action on environmental problems cannot be understood in any straightforward sense. With respect to this issue, it is highly problematic to attribute to individual interventions "direct material effects" that are "visible" and "enduring," to use the terms in which Macnaghten characterized "credible" forms of environmental citizenship. Climate change campaigns that promote energy saving in the home do involve attempts to make such effects more tangible, for instance, by providing calculations of the number of tons of CO_2 in emission reductions that would be accomplished if a certain percentage of Londoners would "turn the thermostat down one degree." But the effects this would have on climate change generally remain shrouded in silence. Still, it seems a mistake to conclude from this that the project of establishing material or physical connections with the issue of climate change is only marginally relevant to the campaigns that focus on greening the home. It precisely seems to be one of the distinctive affordances of household devices, in the context of climate change, that they somehow enable people to "relate" to the issue via material and physical linkages, that is, via the technologies that connect them with energy infrastructures. However, it seems difficult to account for this if we understand the recent focus on the home as a site for civic involvement in climate change as principally an attempt to bridge the *phenomenological* gap between citizens and the environment.

The significance of domestic technologies, as material or physical objects, does become clearer when we consider the recent turn in climate change campaigns to the domestic setting in a broader political and economic context. Thus climate change today serves as a major justification for large-scale projects of regulatory, financial, and industrial restructuring that are to facilitate the transition to a "green energy economy." In this context, the home has been singled out as a major location in which this transition is to be undertaken. Thus around the same time that energy companies and governmental bodies launched climate change awareness campaigns centered on the home, the new U.K. prime minister, Gordon Brown, announced that the building of carbon-neutral homes would be a central policy objective of his government.[6] In this context, publicity campaigns that articulate the home as a site for civic involvement in climate change can be understood as part of the wider project of "preparing the ground" for a new political–economic regime organized around sustainable energy. More particularly, they can be understood as helping to facilitate the emergence of the "green energy consumer," a subject for which there is an obvious need in the low-carbon economy of the future. Importantly, Elisabeth Shove could argue only a few years ago that the "energy user" did not really exist as such, as few people approached domestic practices in terms of the consumption of energy involved in them, and most did not pay much attention to their electricity and gas bills (Shove 2003). From this perspective, publicity around the "simple steps" that can be taken in the home "to help save energy, and the environment" can be understood as an effort to articulate situations in everyday life where (sustainable) energy consumption takes place and where, accordingly, people may adopt and cultivate a new identity as (green) energy consumers. Domestic technologies like energy-efficient lightbulbs and mobile phone chargers may then be understood as devices that can help to make energy consumption "legible" as part of daily life, providing the means with which the "new" activity of sustainable energy consumption can be performed. These devices then enable people to undertake, simply by installing or unplugging them, their own personal transition to becoming active and responsible subjects under the new sustainable energy regime (Rose 1999; Shove 2007).

Such a widening of perspective brings into view close continuities between environmental awareness campaigns and processes of the material organization of social life. Among others, it suggests that household technologies can be understood as material "extensions" of technologies of publicity. With the aid of publicity media, these devices can be repurposed as civic technologies that practically enable people to adopt the identity of "low-carbon" citizenship. But an approach that focuses on the role of "technologies of citizenship" in the management of political economic regime change also has important limitations insofar as it favors a reductive account of civic involvement in climate change.

Thus such an approach defines citizens as subjects that principally exist in relation to the state, or at least to a political economic regime of "green" governmentality, rather than in relation to issues. That is also to say, the relations that people may seek to establish with an environmental problem like climate change, via the home, here appear as essentially mediated by political economic regimes. Indeed, this issue here seems to matter only to the extent that it is mobilized as a relevant "framing" in political and economic discourses on the transition to the low-carbon economy. Thus consideration of the transition to a green political economy may help us appreciate the significance of material practices for projects of civic involvement in climate change, but it leaves unclear how material connections might mediate people's involvement *in this issue.*

Some authors have sought to develop more constructive accounts of issues as objects of public involvement and the importance of publicity media in this regard. Thus Andrew Barry has argued that the mediatization of environmental issues, as, for instance, air pollution in west London, presents an opportunity for inventive forms of civic engagement. He has pointed out that public reporting, however much it may be geared to the stabilization of problems, solutions, and identities, enables third parties, such as activist groups, to use the media to open up these stable definitions for criticism (Barry 2005). Importantly, Barry has drawn attention to the fact that such practices of public contestation may themselves take the form of material practices, as in the case of roadblocks undertaken by activist groups in southern England in the late 1990s (Barry 2005). He describes how in situ protests in this case became media events and, under these conditions, the protestors could make use of the material setting (road construction sites surrounded by English landscape) as an instrument for the articulation of environmental concerns. In a review of Barry's work, Michel Callon has further elaborated this point by suggesting that the articulation of nonhuman entities in publicity media, like air quality, presents an important enabling condition for public debate about environmental issues (Callon 2004). In the view of Barry and Callon, then, the publicization of physical and material entities in the media should not only be understood in terms of institutional efforts to "govern," though this is certainly an important aspect of it. It also presents a condition of possibility for public involvement in issues to the extent that the publicization of entities enables people to relate to them in their capacity as members of the "public." Barry and Callon thus open up a constructive approach to the mediatization of material practices in the context of environmental politics. However, it is also striking that in conceptualizing public involvement, Callon and Barry principally use discursive metaphors, characterizing it in terms of debate and contestation. In this respect, their studies cannot tell us to what extent material practices can themselves be understood as forms of public involvement in issues.

From Clandestinity to Publicity: A "Coming Out" for the Politics of Technology

That it is difficult to account for material practices as sites of public involvement in issues may have to do with a broader conceptual problem. It may have to do with the fact that, in recent social and political theory, the relations between politics and material practice have been understood in a way that *excludes* consideration of publicity. Under the influence of Michel Foucault, sociologists have from the late 1980s onward turned their attention to the affordances of material arrangements and, in particular, technologies for the pursuit of political projects (Winner 1986; Latour 1992). An important assumption of this line of work has been that the success of material politics partly depends on a circumstance that is almost the opposite of publicization: the fact that things are not generally recognized as significant "agents" of politics. Thus Langdon Winner's seminal text on "the politics of technology" focused on a relatively unassuming aspect of the built environment, traffic bridges, which were constructed in the 1930s on Long Island, where they prevented buses, and thereby black people, from visiting the peninsula. In Winner's account, the fact that few people would suspect bridges of pursuing a "racist" politics comes to the fore as an important precondition for the production of political effects by material means. And subsequent work in the sociology of technology, as that of Bruno Latour (1992) on speed bumps, has equally suggested that the ability of technologies and material artifacts to intervene "politically" in the world depends on their relative unobtrusiveness, on their clandestine status, as active components of social and political arrangements.[7] This assumption is also present in studies of the role of material entities in the mediation of *civic* relations, and more particularly, of energy technologies as devices of citizenship. Madeleine Akrich has suggested that the installation of electricity meters in homes in Côte d'Ivoire in the 1980s should be understood as an attempt to foster citizenship. As the government of Côte d'Ivoire had few resources at its disposal for involving people as citizens in the state, the national electricity grid became an important means for implicating people in the political order. Thus, in the very process in which people were enlisted as consumers of electricity, Akrich writes, they were also enrolled as subjects of a nation-state in the making (Akrich 1992). Thus, in Akrich's study, energy technology acquires an important role as a technology of citizenship in a context in which publicity media are absent.

This preoccupation, in recent literature on the politics of technology, with the clandestine production of political effects may help to clarify why it is difficult to account for the forms it takes under conditions of publicity. Authors like Winner, Latour, and Akrich have done crucial work in elucidating how material practices may serve as sites for political intervention, but their studies exclude consideration of the role of publicity in this respect. As they conceive of material politics as a form of subpolitics that plays itself out below the threshold of public perception, their approaches do not help to make it clear how to conceive of the role of publicity in this context. One could say that the materialization of politics, in work on the politics of technology, coincides with its evasion from sites of publicity, its *depublicization.* A related problem with the account of the politics of technology as subpolitics, in this regard, is that it does not consider material politics in relation to *democracy*; rather, this line of work continues to feed suspicions that a politics pursued by material means presents a non-, post-, or even antidemocratic form of politics as it is clearly out of line with familiar understandings of democratic politics as involving collective processes of will formation, institutional evaluation, and public debate. Importantly, attempts to address such suspicions in proposals for the "democratization" of the politics of technology do not necessarily dislodge the association of material politics with clandestine, not quite kosher, forms of intervention. This is because such proposals have mostly taken the form of procedural designs for events of "stakeholder participation" and "public debate" concerning technology, and as such they suggest that democratization of the politics of technology requires its displacement away from material practices to settings of discursive engagement (Marres 2005; de Vries 2007). In presenting discursive processes of negotiation and debate as the principal conditions for democracy, such proposals then leave the understanding of material politics itself to a large extent untouched. Interestingly, however, a number of authors have more recently begun to address questions of the place of materiality, and the nonhuman world more broadly, in democracy.

Sociologists, geographers, and political theorists have over the last years drawn attention to the fact that modern understandings of politics and democracy limit participation in it to human actors (Latour 1993; Mol 1999; Whatmore 2002; Bennett 2001). Interested in the potential gains of redressing this imbalance, these authors have explored the possibility of reconfiguring concepts of political community to include nonhuman entities. Perhaps most important, they have proposed the concept of "heterogeneous assemblages" as a way of taking into account that physical and material entities may figure as active elements in political configurations. In adopting concepts like this, these theorists could be said to undertake a "Gestalt switch" from a human-centered conception of community to the notion of configurations of human and nonhuman entities as a notable site where politics plays itself out. This shift has the potential to recast many of the questions of political theory (Latour 2004a; Mol 2002; Bennett 2001). It suspends the belief that nonhumans can be contained in essentially passive categories like the "topics" of political debate and the "means" and "objects" of political action; that is, it presents a break with the instrumentalist assumption that insofar as politics is concerned, nonhuman entities can be principally characterized in terms of their susceptibility, or lack thereof, to manipulation by human actors, in their role of participants in debate, and decision and policy makers. Focusing on heterogeneous assemblages is, then, a way of recognizing that nonhuman entities are capable of actively making a difference to the organization of social, political, and economic arrangements. For this reason, these authors propose, they must be taken into account as active elements in these arrangements. Importantly, as this line of work is concerned with the "coming out" of nonhumans as significant members of social and political formations, it encourages us to consider how nonhumans are articulated as such in the realm of publicity media. However, this certainly does not mean that the association of material politics with subpolitics is ruptured in this line of work.

Thus some students of heterogeneous assemblages have expressed positive appreciation for the covert status that nonhumans enjoy in the world of politics. Thus Hinchliffe et al. (2005) have emphasized that the relative clandestinity of nonhumans in the political realm does not only signal their undesirable "marginalization" but has affordances as well. Perhaps most important, it opens up a space for situated involvements with these entities as singular beings. As nonhumans prove resistant to assimilation into preexisting definitions of either the subjects or the objects of policy, they must be engaged in their idiosyncrasy. The political affordances of clandestinity have also been stressed with respect to the role of physical entities in democratic politics. Thus, drawing on the work of Jacques Ranciere, Jane Bennett (2005) has argued that the location of nonhumans below the threshold of discourse enables them to interfere surprisingly in political force fields, an event that in her view is crucial to democracy. Work on heterogeneous assemblages, then, does not necessarily dissolve the notion that the politics of nonhumans operates primarily on the subpolitical level. Indeed, it suggests that publicization of nonhuman entities may hamper rather than amplify their capacities to produce political effects. This position raises some difficult questions such as whether the commitment to recognize nonhumans as constitutive elements of social and political worlds does not require some kind of commitment to publicity as one of the principal instruments to bring such recognition about; that is, one can ask whether a positive appreciation of heterogeneous polities, on theoretical grounds, does or should not imply an appreciation of the practical means by which the "coming out" of heterogeneous assemblages can be realized, that is, publicity media?[8]

However this may be, other work in this area *has* made the public articulation of assemblages its explicit concern. Thus, in work that has close affinities with that of Barry and Callon discussed earlier, Bruno Latour has proposed the notion of "matter of concern" to describe the emergence of issues in which human and nonhuman entities prove to be intimately entangled (Latour 2004b). Drawing on the example of the Columbus space shuttle disaster in 2003, Latour shows how in this event a tangled "object" was articulated, with the aid of live media, which included an impressively wide range of elements, from insurance companies to the gods that live in the heavens. His account of this process of public articulation emphasizes that as these divergent but entangled elements came into public view, a multiplicity of issues became subject to scrutiny all at once: the scientific methods of monitoring spacecrafts during flight, the economic question of the costs and gains of the implementation of safety measures, the moral issue of whether individuals should be held responsible for the accident, and the religious concern of how one relates to the gods in the case of human deaths. Importantly, this means that matters of public concern in Latour's account are no pure entities that would fit one rather than another concept of the common good, but rather present messy bundles of things and questions, of which it is still to be seen with which understanding of "morality" or "science" they could be made to comply. However, Latour's suggestion that the emergence of a matter of concern involves the simultaneous "activation" of scientific, economic, political, moral, and religious issues raises the question of what exactly is specific about the mode of entanglement he calls a matter of concern, and about our way of relating to it. Importantly, Latour highlights that our relation to these matters—whether it takes the form of attention, interest, or involvement—should be understood in terms of *attachment;* that is, to be concerned, in his view, is a matter of being noticeably entangled with entities that are at risk and that may well put one's mode of existence at risk. However, Latour's account of matters of concern does not sharply distinguish a mode of attachment that is characteristic of publics or citizens, as opposed to persons, in their capacity of scientific, mortal, economic, or private beings.

In this way, Latour's notion of matters of concern, like other studies of the role of nonhumans in politics mentioned earlier, to an extent leaves undiscussed the specific features of heterogeneous assemblages as objects of publicity, and of public involvement. In other work, Latour does develop a conception of the public, which he derives from the political theory of the American pragmatist philosopher John Dewey (Latour 2001). In Dewey's work, Latour finds an important precedent for a definition of the public as an attached being, whose concerns derive from the entanglements of everyday life. That is also to say, Latour turns to Dewey for an alternative to the republican conception of the public as consisting of actors who are detached from the concerns of everyday life and concern themselves with matters of general, as opposed to particular, concern. Dewey's political theory dissolves the notion of two distinct domains, the public and the private, and indeed, in doing so, he directs attention to something that we can retrospectively recognize as "heterogeneous assemblages," as one of the key sites in which political relations are constituted. Dewey's work thus presents a crucial point of reference for those with an interest in developing a constructive account of the role of nonhuman entities in politics (Bennett 2005; Marres 2005; Stengers 2006; Dijstelbloem 2006). However, his political theory may also be a helpful guide in exploring the more specific question of how to distinguish "public involvement," as a mode of relating to heterogeneous assemblages, from other modes—a question that becomes crucial when we recognize that public involvement practices are performed *in media res* and not only in dedicated domains distinct from social life. Thus I would like to suggest here that Dewey's theory of the public can productively inform an account of material practices as sites of public involvement and of the importance of publicity media in this respect. Moreover, such a reading of Dewey may also help to make it clearer what the distinctive contribution of studies of heterogeneous assemblages to the study of public involvement in politics consists of.

John Dewey's Heterogeneous Public

Those with an interest in the roles of nonhumans in politics are certainly not the only ones to have turned to the political theory of John Dewey in recent years. A wide range of authors in contemporary political theory draw on his work for a variety of purposes: to expand and strengthen the deliberative conception of democracy as anchored in public debate (Festenstein 1997), to establish the importance of technological innovation as an occasion for public participation experiments (Keulartz et al. 2002), or to conceptualize minority politics as a practice grounded in experience and not doctrine (Glaude 2007). However, recent readings of Dewey's political theory, and in particular of his theory of the public, in the light of work on the role of nonhumans in politics offer a distinctive interpretation. As they highlight that Dewey conceived of the "public" as constituted by materially and physically entangled actors, they break with an assumption shared by many interpreters of his work, namely, that to participate in a public is principally a matter of participating in discursive exchange (Marres 2007). As will become clearer later, such differences among interpretations can, to a great extent, be accounted for in terms of the different books, or even passages of books, on which different interpretations focus. Thus Dewey (1927) introduced his "heterogeneous public" in the first chapters of *The Public and Its Problems.* The book opens with the outline of a speculative history of the emergence of political formations, and befittingly, Dewey develops an account of the public as emerging from the ever-shifting relations between humans and nonhumans as part of this historical exposé. The public, Dewey argues here, can be defined as a particular type of distribution of the consequences of human action: "the public consists of all those who are affected by the indirect consequences of transactions, to such an extent that it is deemed necessary to have those consequences systematically cared for" (15).

A lot is packed into this brief definition, with the noteworthy nonhumans hiding, for the moment, in the notion of the consequences of action that actors are harmfully, or at least disturbingly, affected by. But it seems most useful to begin picking Dewey's concept of the public apart by considering how it resists reduction to another familiar understanding of the public, even if it bears similarities to it. Thus Dewey's public can in some respects be understood as an elaboration of the liberal ideal that says the interference of government in private affairs should be limited to those situations in which persons suffer as a consequence of the actions of third parties. Dewey, one could say, transformed this regulative principle, designed to *limit* state involvement in "private matters," into a constructive principle that can account for the empirical process of the formation of publics—and that indeed *extends* the range and number of empirically existing publics in comparison to other definitions. Dewey does this by concentrating on the type of consequences that in his view call the particular figure of the public into existence. These are consequences that are (1) harmful or even "evil" (Dewey 1927, 17), (2) indirect, and (3) extensive and enduring. And in focusing on these effects, Dewey dismantles the particular opposition between the public and the private that is central to this liberal project of restricting "incursions" of the former into the latter. (That is also to say, Dewey was at least as much concerned with questioning liberal as republican concepts of the public.)

Thus, rather than presenting us with two domains, one in which we deal with personal matters and another in which we deal with common affairs, Dewey presents a world in which actions continuously produce actor groupings by way of their *effects,* some of which will go by the name of "public." Dewey singles out two critical features of such public-generating effects: he distinguishes first between consequences that are direct and can thus be controlled by those involved in their production and those that cannot, and second between consequences that are erratic and somehow can be accommodated as part of social life and those that produce enduring harm. Consequences that have the latter features generate publics. Dewey thus exhaustively defines the public in terms of a particular chain of effects, which can be differentiated from other such chains, but both of which proliferate across one and the same worlds. Thus, as he situates the public in effects and affects that are continuously produced as part of daily life, the notion that the public refers to a domain that exists apart from private worlds loses much of its sense. Moreover, Dewey is clear that the type of consequences that produce publics can be expected to be generated everywhere all the time. He inferred from this that problems of democracy are in fact not likely to stem from a shortage of publics but rather from their radical *multiplication* and excess (126). Thus, far from limiting the breadth and scope of the public, in line with the classic liberal objective, Dewey radically extends them. Furthermore, Dewey could also be said not just to redefine the opposition between public and private but to replace it with a different one; that is, in some respects, it makes little sense to oppose Dewey's public to the "private." In Dewey's account, the event of "incursion," that is, when people experience harmful indirect effects, does not present a situation in which private actors are threatened by an external force. It rather transforms social actors who more or less "routinely" went about their daily lives into a public that must take it upon itself to organize into an external force (vis-à-vis the actions that must be intervened in, if harmful effects are to be addressed). Thus, rather than the intrusion of the public into the private, the central event of his account of the public appears to be the rupture of habitual ways of doing, which results in the formation of a public. In Dewey's account, the state of being harmfully affected by events beyond actors' control requires the formation of a collective agency and, more generally, the need to get involved in something like politics. Dewey's theory of the public could thus be said to replace the opposition between the private and public domain with the notion of a shift from working social routines to their disruption.

Importantly, Dewey's emphasis on the disruptive events in which publics come into existence also sets his account apart from other consequentialist approaches to morality and politics, such as utilitarianism. Dewey *did* follow utilitarianism by concentrating on consequences, but he certainly did not subscribe to its conception of politics as principally concerned with the maximalization of "agreeable" consequences of action and the prevention of disagreeable ones. In other work, Dewey expressed great appreciation for the fact that utilitarianism, by focusing on the consequences of action, was able to recognize the "empirical character" of morality and politics. But he was extremely critical of the utilitarianist notion that it is possible to *calculate* future consequences of action and also of the distinction between means and ends on which such a calculative approach is predicated (Dewey 1922); that is, Dewey rejected the utilitarianist definition of politics and morality as concerned with the determination of the proper means that will help to realize specifiable desired ends because he could not accept the instrumentalist's carving up of the world into means and ends that this implied. He criticized the role that the means–end distinction was made to play in politics and morality by utilitarianists because of the way in which it precluded recognition of the fact that things designated as "means" are likely to have consequences that are not included among its "ends." To approach such things as "mere means" to "certain ends only" for Dewey presented a disingenuous justification for excluding these consequences from consideration (222–27). (This is also to say, while Dewey called his own philosophy instrumentalism, it is clear that he meant something quite different than the utilitarian brand of "means-ism.") In *The Public and Its Problems,* this criticism of the notion of "mere means" also returns, when Dewey highlights the relevance for politics of the situation in which things that are designed to function as means of human action produce unanticipated effects. Indeed, it is in his discussion of this situation that Dewey comes to recognize the formative influence of nonhuman entities on the organization of publics. Thus, in specifying the conditions in which publics come about, Dewey directs attention to the tendency of technological means not only to produce consequences that cannot be classified among those that are desired but also to produce new types of consequences: "industry and invention in technology, for example, create means which alter the modes of associated behavior and which radically change the quantity, character, and place of impact of their indirect consequences" (30).

Passages like this help to make clear how a definition of the public in terms of a particular type of consequence involves recognition of the role of nonhuman entities in the formation of publics. First, it highlights Dewey's conviction that it is unhelpful to define nonhuman entities as mere means in the political context. Technologies, substances, and objects play a crucial role in the formation of publics because they actively participate in the production of the consequences that call publics into existence. Second, it also shows that Dewey includes, or even privileges, among the consequences that produce publics material and physical effects that have to do with activities like manufacture, transport, and communication. Indeed, one of Dewey's aims in *The Public and Its Problems* is to direct attention to changes in "the material conditions of life" as a crucial occasion for the formation of publics (Dewey 1927, 44) and for the development of democratic societies more broadly. Thus, by defining the public in terms of adversely affected actors, Dewey suggests that we should look for a distinctively public mode of association not, in first instance, in features like shared membership in clubs and other social associations or in shared discourses. We should rather focus on the social fact of the joint implication of actors in the infrastructures, technological, natural, and otherwise, that sustain social life. In this respect, it is important to note that Dewey's *The Public and Its Problems* is for a large part concerned with problems of democracy in technological societies. By defining the public as he does, Dewey in effect breaks with the tendency in political theory to model the public on preindustrial communities of either the aristocratic or agricultural variety. Dewey does not, at least not initially, mold his public after a particular social community, be it the community of notables (citizens of the polis) or the New England village (meeting in the town hall). He opens up the concept to the ever-shifting and complex interdependencies that are characteristic of industrial societies.

However, Dewey's emphasis on the material and physical connections by which publics, in his view, are held together does not entail a disregard for the importance of discourse. Interestingly, and it could make sense to call this Dewey's genius, he makes it clear that the "material public" conceptualized by him needs *more rather than less publicity* to sustain itself, compared with communities that are principally held together by discursive or social bonds. The notion of publics called into existence by material effects opens up the possibility that publics proliferate in the absence, or below the surface, of the usual support systems that these other publics require: a shared way of life, discourse, institutions, assembly spaces, publicity media. But in *The Public and Its Problems,* Dewey utterly refuses to view this possibility of what could perhaps be called the subpolitical mode of existence of the public in a positive light. Instead he argues that publics that configure around the harmful consequences of human action depend for their survival and their "effectivity" on publicity media. In Dewey's view, material publics are condemned to lead only an inchoate, obscure, staggering, and unstable existence, as long as they remain aloof from the symbolic circulations facilitated by publicity media, and this for at least two reasons. First, the consequences that call publics into existence are unlikely to be recognizable as such if they are not documented in information media. This is because these consequences, being indirect, are likely to transgress the boundaries of existing social groupings. In industrial societies, moreover, with their longer and more complex associative chains, these effects also tend to be extensive, connecting actors that are separated from one another by long distances. And they should be expected to contain an element of novelty. From this Dewey infers that, in the absence of attempts to trace indirect and harmful consequences with the aid of information technologies, the formation of a public is likely to go unobserved (Dewey 1927, 177). Second, it also follows that a public is unlikely to recognize itself as long as the effects that call it into being are not made *widely* observable. As Deweyian publics are not likely to map onto existing social groupings, they should be expected to consist of strangers who do not have at their disposal shared locations, vocabularies, and habits for the resolution of common problems (Warner 2002; Dobson and Bell 2006). From this Dewey concluded that if publics are to "recognize themselves," platforms for the wide and open-ended circulation of information concerning consequences must be in place. And more generally, he argued that an extensive and developed communication infrastructure is a central requirement for the organization of publics. It thus seems fair to say that he was not in the least seduced by the political possibilities inherent in an exclusively subpolitical form of organization for which his concept of material publics can seem to allow. Publicity, in his view, was an absolutely necessary condition for the endurance or sustainability of material publics, that is, if they were to develop capacities to "hold themselves," as well as for any possible effective action on the part of the public.

This brings us to a point at which Dewey's theory of the public at once touches most closely on questions in contemporary political theory regarding materiality and politics and begins to recede from them. On one hand, Dewey's concept of material publics—and, perhaps especially, his claim that they depend for their sustenance and effectivity on the publicization of the effects that call them into existence—seems to contain the seeds of answers to these questions. It suggests how a special combination of material effects, intimate affectedness, and mediatization comes together in the figure of the public. It distinguishes, within the wider field of subpolitical formations involving things, humans, and environments, a particularly problematic type of entanglement of humans and nonhumans, simply and elegantly called "public," the articulation of which requires publicity. I will further discuss subsequently how these Deweyian concepts can help to address a number of conceptual complications regarding materiality and citizenship. However, on the other hand, it should also be noted that Dewey's claims about the dependency of material publics on publicity media present the point in *The Public and Its Problems* at which his argument starts to be less and less relevant to these complications. Indeed, it seems that, partly because of his preoccupation with the communicative dimension of the public, Dewey was unable to fully appreciate its heterogeneous character. Thus, once Dewey has established the importance of informational and communicative practices in this book, his account begins to move away from the idea that the public is constituted by human as well as nonhuman entities. Indeed, it subsequently becomes clear that Dewey, in certain respects, remains firmly committed to a humanist understanding of the public.[9] Thus, in *The Public and Its Problems*, he eventually comes to define social and political groupings in terms of the "conjoint activity of humans." Moreover, in doing so, he makes the demarcationist move of distinguishing human communities from nonhuman ones rather than continuing to explore their mutual imbrications. He places great emphasis on humans' exclusive mastery of language and symbolic communication. Dewey thus ultimately came to define political groupings in terms of associations among distinctively human beings, and the notion that nonhuman entities make a difference to the political formations they help to constitute disappears from his argument. That is also to say, Dewey at no point addresses the question whether the participation of nonhumans in the public has consequences for the forms that practices of publicity may take. However, in the context of the recent turn to material practices as sites of citizenship, this seems to be one of the central questions raised by Dewey's theory of the public: once we recognize that publics are heterogeneously constituted, must not publicity itself—the process in which publics come to "recognize themselves" and somehow acquire the capacity to act—be rethought along materialist lines?

Before further discussing what inferences can be made from Dewey's political theory regarding this question, I want to briefly point out that, in other work, Dewey did emphasize the special affordances of objects as mediators of normative engagement. Thus, in his *Theory of Valuation,* Dewey (1939) developed a moral theory that, without much exaggeration, can be characterized as object oriented. Here he proposes that values, as well as the desires and interests that guide their articulation, are first and foremost attached to objects, and that they are most productively defined in terms of those objects. Interestingly, *Theory of Valuation* was published in the famous series edited by the Viennese logical positivists, *The Encyclopedia of Unified Science*, and its general argument can be understood in relation to the commitments of logical positivism. Thus one could say that Dewey, in this little book, presents an alternative to the positivist project of purifying the domain of factual truth and excluding from it anything "subjective," which logical positivists famously equated with thoughts and feelings that are merely "confused." In sharp contrast to this, Dewey proposed to include affectations, whether confused or not, in the objective realm. As he puts it elsewhere, "such things as lack and need, conflict and clash, desire and effort, loss and satisfaction [must be] referred to reality" (Dewey 1908, 124). In *Theory of Valuation*, he specifies this general claim by suggesting that values, desires, and interests must be appreciated as aspects of "objective situations." These normativities in his view first and foremost connote "an active relation to the environment" (Dewey 1939, 16); they must be stated "in terms of the objects and events that give rise to [them]" (16); and the tendency to define them as "something merely personal" must be resisted (16). For Dewey, the content of "values" is then best conceptualized in terms of the specific situational objects to which people attach them, and he suggests that processes of valuation should themselves be understood as processes in which the worth of this type of objects becomes clear. Thus, regarding interests, Dewey states, "When [they] are examined in their concrete makeup in relation to their place in some situation, it is plain that everything depends on the objects involved in them" (18). This is not the place to examine this moral theory in detail. But I hope that this brief sketch is enough to make it clear that Dewey's philosophy contains further conceptual elements to help account for, if not material, then at least object-oriented practices as sites of normative engagement. One last point, which Dewey derives from his object-oriented conception of values, seems especially relevant in this respect. This conception led him to express a strong commitment to *action* as the appropriate register for the expression of value. As he put it, "the measure of a value a person attaches to a thing is not what he *says* about its preciousness, but the care he devotes to obtaining and using the means without which it cannot be attained" (27).

Thus Dewey's object-oriented understanding of values led him to foreground practical efforts to obtain valued things as a privileged mode of moral action. He was critical, certainly not of moral discourses in general, but of a particular tendency in the expression of moral sentiment, one that ends up "merely wishing" that "things were different" (Dewey 1939, 15). The problem with wishing, Dewey points out elsewhere, is that it may all too easily entail a disregard of the issues people are confronted by, as people "tend to dislike what is unpleasant and so to sheer off from an adequate notice of that which is especially annoying" (Dewey 1933, 137). For Dewey, valuation crucially involved an acceptance of the practical costs of engaging what he famously called problematic situations, those involving "lack and need, conflict and clash, desire and effort, loss and satisfaction."

The Particularity of Material Politics

Dewey's philosophy is a productive one also in the sense that there can easily seem to be no end to the connections that can be explored between his various concepts. But I hope that the preceding goes some way toward making clear how his theory of the public can help to elucidate contemporary questions about the relations between materiality and publicity. It can do so not in the least because it opens up a perspective on the role of materiality in politics that breaks with the tendency, present in post-Foucauldian work on the subject, to understand this role as antithetical to publicity. Crucially, to adopt the Deweyian concept of "material publics" does not imply a rejection of the association of material politics with clandestinity, with the idea of a force at work below the surface of publicized reality; rather, Dewey's *The Public and Its Problems* proposes that among the many different types of human and nonhuman entanglements that exist, there is a *distinctive type*, simply and elegantly called public, of which the articulation requires publicity. This also means that Dewey's work raises a slightly different question than the one on which authors concerned with the politics of technology have focused. The latter were interested in the question whether assemblages of humans and nonhumans can be ascribed a politics *generally speaking* (Harbers 2005). By contrast, Dewey directs attention to a specific type of assemblage to describe how heterogeneous assemblages may become politically charged. Because of this, Dewey's theory of the public also opens up an alternative interpretation of the idea of subpolitics. The latter notion has been criticized for suggesting that politics happens everywhere all the time, and it has therefore been said to contribute to the dismantling of the concept of politics (de Vries 2007). Such a critique gives rise to the temptation to confine the politics of technology to a particular institutional domain, where the specificity of politics can be safeguarded. In contrast to this, Dewey's concept of the public suggests that it is certainly not necessary to relegate politics to a separate domain, if the point is to acknowledge its specificity. Indeed, the project of restricting politics to an institutional domain is precisely the kind of classic liberal move that Dewey's concept of the *infrapublic* is designed to undermine. This concept captures the specificity of political relations by directing attention to a particular mode of association among social actors: that of being jointly affected by actions beyond their control.

Furthermore, though Dewey rejects the understanding of the public in terms of a separate domain, he nevertheless emphasizes that there is something distinctive about being implicated in heterogeneous assemblages *as a member of the public*. In his account, the public's position is marked by a special combination of being both an insider and an outsider to public affairs. Thus Deweyian publics are internal to public affairs to the extent that they are *intimately* affected by social problems, which put their livelihood, in the broad sense, at stake. But they occupy an external position to the extent that the sources of social problems are beyond their reach and control, and, we should add, so are the resources required to address them. In this way, Dewey makes it clear that the public's mode of involvement in social problems should be differentiated from those of social actors and other particular actors like professionals. This distinction is not always made in studies of heterogeneous assemblages, which tend to focus on the situated involvements of various social actors in them. Also relevant in this respect is that it is Dewey's insight in the singularity of the public's position that subsequently leads him to recognize the importance of publicity media. The location of the public as both an insider and outsider to social problems raises the question of how such a position can be sustained, and Dewey's answer is to point at publicity media. He arrives at the intriguing position that the existence of material publics, which are called into being by harmful consequences, depends at least partly on their articulation in media. Because the effects that call publics into existence are so obscure—that is, precisely because they present "clandestine" formations—Dewey suggests they can only exist coherently in these distributed, formal, artificial platforms. As I mentioned, Dewey's recognition of the importance of publicity media led him away from his earlier concerns with the materialities that mediate the formation of publics. But his initial account of their role does suggest an approach to material practices as sites for public involvement. Taking a cue from *Theory of Valuation*, in which Dewey argued in favor of practical interventions as a mode of normative involvement in things, we can ask whether this argument cannot be extended to the mode of involvement characteristic of the public. Thus, in the light of Dewey's definition of the public in terms of its state of "being affected" by public affairs, object-oriented practices appear to have special affordances. *The Public and Its Problems* makes it clear that this state of "affectedness" cannot be adequately understood in factual terms only but also refers to the affective states of being touched, implicated, and indeed moved in the sense of being mobilized by public affairs. So how is the state of affectedness that is characteristic of the public performed and made productive? Specific objects may have crucial enabling features in this respect.

But before saying a final word about the affordances of material things for the performance of public involvement in issues, I would like to point out that studies of heterogeneous assemblages in their turn also suggest a particular elaboration of Dewey's theory of the public. This is because these studies have a particular way of dealing with critiques of "naive objectivism," to which Dewey's pragmatism has been subjected. They accommodate these critiques without letting go of the object-oriented approach that is characteristic of Dewey's philosophy, as has been rather more customary. Thus Dewey's political theory has been called historically dated, on the ground that his objectivist approach to democracy can no longer be maintained, for historical, epistemic, and political reasons. Sheldon Wolin (2004) and Yaron Ezrahi (1990) have emphasized that Dewey promoted a scientific approach to democracy that aimed to transform politics and morality into objective practices, dedicated to tracing, documenting, and remedying "harmful" consequences, rather than to subjective processes of will formation and making value judgments. Such a characterization makes Dewey's pragmatism seem more utilitarian and positivistic than is perhaps justified. But it is certainly not entirely wrong. Wolin has described how Dewey's problem-centric understanding of democracy became problematized historically: the adoption of a problem-centric approach by progressive U.S. administrations after the Second World War did not result in the type of enlightened, participatory form of rule to which Dewey was committed (Wolin 2004, 518–19). Thus Dewey's object-oriented politics, Wolin suggests, came down in practice to a form of technocratic government that idealized expert-driven forms of policy making dedicated to narrowly defined ideals of "problem solving." Ezrahi has pointed out that Dewey's objectivist understanding of politics also became problematic toward the end of the twentieth century, epistemically speaking. His belief in the traceability of "harmful" consequences, he notes, involves a commitment to an empiricist ideal of accountability: that is, a belief in the possibility of documenting events and "locating the trouble" without getting caught up in confusing complexities involving interests, obscure motives, and political games of assigning blame. This kind of empiricism has become deeply problematic, if not untenable, Ezrahi points out, as the constructivist commitment to recognize the influence of "paradigms" and "frames" on the formulation of facts has become widely adopted. Finally, both Ezrahi and Wolin have pointed at the *political* impossibility of Dewey's objectivist ideal of democracy. As Wolin puts it, democracy inevitably involves the clashing of views and interests and upheavals having to do with struggles for influence, and Dewey's scientific understanding of democracy failed to make room for such events (Wolin 2004).

As mentioned, one possible response to these critiques is to point out that Dewey was not the utilitarian or positivist that he is sometimes taken for, designations that the preceding critiques perhaps do not do enough to dispel. However, it seems equally important to recognize that the preceding "problematizations" of Dewey's philosophy tend to result in a stalling, or even a reversal, of the objectivist turn that he proposed. Thus Sheldon Wolin concludes his essay on the fate of Dewey's scientific ideal of democracy in the late twentieth century by advocating a return to the ideal of solidarity, as it was expressed in the protest movements of the 1960s and 1970s. In this context, the attempts of students of heterogeneous assemblages to adapt Dewey's political theory to their purposes present a clear alternative. This line of work has been committed to demonstrating that it is possible to address critiques of positivism and utilitarianism, while further radicalizing the object-oriented approaches that these schools of thought opened up. Thus the notion of heterogeneous assemblages has been developed as part of a broader critique of instrumentalism: it is all about recognizing the fragility, volatility, and recalcitrance of entities, both human and nonhuman, in the face of attempts to define them as means toward pregiven ends. However, in this case, the critique of instrumentalism does not lead to less but rather more attention being paid to the capacities of objects to mediate political relations.

To Conclude

But what about the home as a site of public involvement in climate change? Dewey's theory of the public, when read through the lens of recent studies of heterogeneous assemblages, suggests a particular approach to the role of domestic technologies in this regard, namely, to consider them as "devices of affectedness." We then enrich Dewey's definition of the public as held together by the indirect and intimate connections that make up social problems with a decidedly postinstrumentalist emphasis on the active role of things in the mediation of political relations. From such a perspective, energy technologies in the home, like thermostats and water cookers, may perhaps be ascribed special affordances for the performance of the specific mode of involvement in social problems that is characteristic of the public, that of being both intimately and externally affected by issues. It is then certainly not impossible that the little act of "turning down the thermostat" deserves appreciation as a more or less successful attempt, not to save the planet, but to transform the state of being affected by the "impossible" issue of climate change into a viable practice. In the light of the various critiques of instrumentalism discussed earlier, it is clear that the affordances of domestic settings for the articulation of the material and physical modes of being implicated in climate change cannot possibly be approached in the register of facticity as given; that is, the capacities of domestic energy technologies to mediate involvement in climate change can only be understood, to use Dewey's vocabulary, as a situational achievement. The material and technological arrangements that make up homes must then be examined further, if we are to determine their relative capacities for dramatizing connections between practices "in here" and changing climates "out there." Perhaps it is not completely anachronistic to suggest that Dewey has made it clear that the capacity of the home to function as a device of issue affectedness depends crucially on the articulation of connections, between domestic practices and issues out there, in publicity media. Whether recent publicity campaigns, with their focus on a limited number of feasible, stereotypical interventions—washing at low temperatures, unplugging mobile phone chargers—succeeded in mediating affective relations with climate change, and thus in bringing the issue home, must remain an open question here. However, to leave this question open is to consider it a real possibility that the endless repetition of suggestions of "what you can do" activates a different, more classic function of the home:

that of a machine of disaffectedness, which has the special affordance of providing shelter against the lures and risks of public life, not the least of which seems to be hyperbole and thereby the loss of connection with its objects.

Notes

1 The strong language was used by British Gas as part of its advertising campaign "Make It Greener Where You Are," http://www.makeitgreenernow.co.uk/.
2 As the campaign "DIY Planet Repairs" of the mayor of London claimed; see "Make Six Small Changes to Help Repair the Planet Says Mayor," press release, Mayor of London, June 6, 2007, http://www.london.gov.uk/view_press_release.jsp?releaseid=12230.
3 The GREENhomes Concierge service, http://www.greenhomesconcierge.co.uk/.
4 "M&S Helps Customers to 'Think Climate' by Relabelling Clothing," press release, Marks and Spencer, April 23, 2007, http://www.marksandspencer.com/gp/browse.html?ie=UTF8&node=55319031&no=51444031&mnSBrand=core&me=A2BO0OYVBKIQJM; see also http://www.together.com/solutions/9.
5 Marketing Plan: Planet DIY Repairs, Mayor of London, May 2007.
6 "Brown Outlines 'Eco Towns' Plan," BBC News, May 13, 2007, http://news.bbc.co.uk/1/hi/uk_politics/6650639.stm.
7 Latour and other actor–network theorists have been criticized in the past for giving too much credit to nonhumans by effectively approaching them as "actors." Such a critique, it seems to me, does not sufficiently appreciate that actor-network theorists tend to limit the agency of nonhumans to "acts" that are precisely not conventionally defined as such.
8 Another way of phrasing this problem is that work on heterogeneous assemblages does not always provide a clear answer to the question of whether these assemblages are best appreciated as polities *by designation only* or whether the shift in perspective it proposes also invites or necessitates appreciation of attempts at the articulation of assemblages as "objects of politics" in social, political, and public settings.

9 This is also suggested by the fact that Dewey excluded natural events from the range of actions that could bring a public into existence. Only human deeds could in his account give rise to a political community. This limitation may have to do with Dewey's understanding of politics in terms of care for consequences, as an intrinsically human capability. But considering the centrality of "harmful indirect effects" to his definition of the public, it is hard to see how Dewey could deem it justified.

References

Akrich, Madeleine. 1992. "The De-scription of Technical Objects." In *Shaping Technology/Building Society: Studies in Sociotechnical Change*, ed. Wiebe E. Bijker and John Law, 205–24. Cambridge, Mass.: MIT Press.

Allan, Stuart, Barbara Adams, and Cynthia Carter, eds. 1999. *Environmental Risks and the Media.* London: Routledge.

Barry, Andrew. 2001. *Political Machines: Governing a Technological Society.* London: Athlone Press.

———. 2005. "The Anti-political Economy." In *The Technological Economy*, ed. Andrew Barry and Don Slater, 84–100. London: Routledge.

Bennett, Jane. 2001. *The Enchantment of Modern Life: Attachments, Crossings, and Ethics.* Princeton, N.J.: Princeton University Press.

———. 2005. "In Parliament with Things." In *Radical Democracy: Politics between Abundance and Lack*, ed. Lars Tonder and Lasse Thomassen, 133–48. Manchester, U.K.: Manchester University Press.

Callon, Michel. 2004. "Europe Wrestling with Technology." *Economy and Society* 1, no. 33: 121–34.

Clarke, Nick, Clive Barnett, Paul Cloke, and Alice Malpass. 2007. "Globalising the Consumer: Doing Politics in an Ethical Register." *Political Geography* 26: 231–49.

de Vries, Gerard. 2007. "What Is Political in Sub-politics? How Aristotle Might Help STS." *Social Studies of Science* 5, no. 37: 781–809.

Dewey, John. 1908. "Does Reality Possess Practical Character?" Repr. in *The Essential Dewey*, vol. 1, *Pragmatism, Education, Democracy*, ed. Larry A. Hickman and Thomas M. Alexander, 124–33. Bloomington: Indiana University Press, 1998.

———. 1922. *Human Nature and Conduct: An Introduction to Social Psychology.* Repr., New York: Cosimo, 2007.

———. 1927. *The Public and Its Problems*. Repr., Athens: Swallow Press/Ohio University Press, 1986.

———. 1933. "How We Think." Repr. in *The Essential Dewey*, vol. 2, *Ethics, Logic, Psychology*, ed. Larry A. Hickman and Thomas M. Alexander, 137–50. Bloomington: Indiana University Press, 1998.

———. 1939. *Theory of Valuation*. Repr. in *International Encyclopedia of Unified Science*, vol. 2, no. 4, ed. Otto Neurath, Rudolf Carnap, and Charles Morris. Chicago: University of Chicago Press, 1955.

Dijstelbloem, Huub. 2006. "De democratie anders: Politieke vernieuwing volgens Dewey en Latour." PhD diss., University of Amsterdam.

Dobson, Andrew, and Derek Bell. 2005. Introduction to *Environmental Citizenship*, ed. Andrew Dobson and Derek Bell. Cambridge, Mass.: MIT Press.

Ezrahi, Yaron. 1990. *The Descent of Icarus: Science and the Transformation of Contemporary Democracy*. Cambridge, Mass.: Harvard University Press.

Festenstein, Matthew. 1997. *Pragmatism and Political Theory*. Oxford: Polity Press.

Glaude, Eddie S. 2007. *In a Shade of Blue: Pragmatism and the Politics of Black America*. Chicago: University of Chicago Press.

Harbers, Hans, ed. 2005. *Inside the Politics of Technology: Agency and Normativity in the Co-production of Technology and Society*. Amsterdam: Amsterdam University Press.

Hinchliffe, Steve, Matthew Kearnes, Monica Degen, and Sarah J. Whatmore. 2005. "Urban Wild Things: A Cosmopolitical Experiment." *Environment and Planning D* 23: 643–58.

Jasanoff, Sheilla. 2001. "Image and Imagination: The Formation of Global Environmental Consciousness." In *Changing the Atmosphere: Expert Knowledge and Environmental Governance*, ed. Clark Miller and Paul N. Edwards, 309–37. Cambridge, Mass.: MIT Press.

Keulartz, Jozef, Michiel Korthals, Maartje Schermer, and Tsjalling Swierstra, eds. 2002. *Pragmatist Ethics for a Technological Culture*. Dordrecht: Kluwer Academic.

Latour, Bruno. 1992. "Where Are the Missing Masses? The Sociology of a Few Mundane Artefacts." In *Shaping Technology/Building Society: Studies in Sociotechnical Change*, ed. Wiebe Bijker and John Law, 225–58. Cambridge, Mass.: MIT Press.

———. 1993. *We Have Never Been Modern*. Trans. C. Porter. Cambridge, Mass.: Harvard University Press.

———. 2001. "From 'Matters of Fact' to 'States of Affairs': Which Protocols for the New Collective Experiments?" Paper prepared for the Darmsdadt Colloquium plenary lecture, March 30. Original document available at http://www.ensmp.fr/~latour/poparticles/poparticle/p095.html.

———. 2004a. *Politics of Nature: How to Bring the Sciences into Democracy.* Cambridge, Mass.: Harvard University Press.

———. 2004b. "Why Has Critique Run Out of Steam? From Matters of Fact to Matters of Concern." *Critical Inquiry* 2, no. 30: 225–48.

Marres, Noortje. 2005. "No Issue, No Public: Democratic Deficits after the Displacement of Politics." PhD diss., University of Amsterdam.

———. 2007. "The Issues Deserve More Credit: Pragmatist Contributions to the Study of Public Involvement in Controversy." *Social Studies of Science* 5: 759–80.

Macnaghten, Phil. 2003. "Embodying the Environment in Everyday Life Practices." *Sociological Review* 51, no. 1: 62–84.

Mol, Annemarie. 1999. "Ontological Politics: A Word and Some Questions." In *Actor Network Theory and After*, ed. John Law and John Hassard, 74–89. Oxford: Blackwell.

———. 2002. *The Body Multiple: Ontology in Medical Practice.* Durham, N.C.: Duke University Press.

Rose, Nikolas. 1999. *Powers of Freedom: Reframing Political Thought.* Cambridge: Cambridge University Press.

Shove, Elizabeth. 2003. *Comfort, Cleanliness, and Convenience: The Social Organization of Normality.* London: Berg.

———. 2007. "Caution: Transitions Ahead." *Environment and Planning A* 39: 763–70.

Stengers, Isabelle. 2006. *La Vierge et le neutrino: les scientifiques dans la tourmente.* Paris: Les empêcheurs de penser en rond/Le Seuil.

Szerszynski, Bronislaw, and Mark Toogood. 1999. "Global Citizenship, the Environment, and the Media." In *Environmental Risks and the Media*, ed. Stuart Allan, Barbara Adams, and Cynthia Carter, 218–28. London: Routledge.

Trumbo, Craig. 1995. "Constructing Climate Change: Claims and Frames in US News Coverage of an Environmental Issue." *Public Understanding of Science* 5, no. 3: 269–83.

Warner, Michael. 2002. *Publics and Counterpublics.* New York: Zone Books.

Weber, Elke. 2006. "Experience-Based and Description-Based Perceptions of Long-Term Risk: Why Global Warming Does Not Scare Us (Yet)." *Climatic Change* 77, nos. 1–2: 103–20.

Whatmore, Sarah J. 2002. *Hybrid Geographies: Natures, Cultures, Spaces.* London: Sage.

Winner, Langdon. 1986. "Do Artifacts Have Politics?" In *The Whale and the Reactor: A Search for Limits in an Age of High Technology,* 19–39. Chicago: Chicago University Press.

Wolin, Sheldon. 2004. *Politics and Vision: Continuity and Innovation in Western Political Thought.* Princeton, N.J.: Princeton University Press.

8 The Political Technology of RU486: Time for the Body and Democracy

ROSALYN DIPROSE

A PUBLIC DEBATE ERUPTED in Australia in late 2005 when, in an unprecedented move, four female senators from across the political party spectrum sponsored a "private members bill" to repeal the federal minister for health's jurisdiction over the licensing of RU486 (the so-called home abortion pill).[1] By February 2006, the newspaper headlines read, "No Pill Has Divided Australia Like RU486 Since the Oral Contraceptive Pill Was Introduced in the 1960s."[2] In contrast to the global spread of the controversy over the contraceptive pill of the 1960s, the eruption of RU486 onto the political scene in 2005–6 was peculiar to Australia. This is partly because the Australian government, in contrast to other comparable democracies, was continuing to block the licensing of the drug despite the fact that surgical abortion has been legally available since the 1970s.[3] But the "RU486 event" was also remarkable in terms of the internal political context. On one hand, abortive agents like RU486 stand out from other therapeutic drugs in being a direct site of government *regulation of the life* of the population. Though the licensing of all other pharmaceuticals had been under the authority of the Therapeutics Goods Administration (TGA) since the Therapeutics Goods Act 1989, in 1996 the newly elected conservative federal government, in one of its first legislative acts since assuming power, transferred the administration of abortive agents to its own jurisdiction. On the other hand, despite this early sign of the spread of this government's authority over life, that RU486 in particular became the focal point of *contestation* of that authority in 2005–6 was surprising. First, the 1996 amendment had had the support of both major political parties in both houses of parliament, and nothing of note had changed about the drug in the intervening ten years. Second, there had been little sign of dissent within parliament to the government's wider legislative agenda up until 2005, and though there had been much dissent outside parliament over a range of government initiatives, the licensing of RU486 had for the most part remained under the public radar. Through eleven years of Liberal-National (conservative) government and increasing authoritarianism,[4] many other issues had more obviously divided the nation (harsh refugee policies, participation in the war in Iraq, draconian antiterrorism legislation, and legislation individualizing and deunionizing the labor market). Yet only RU486 succeeded in mobilizing a wave of defiance of government authority sufficiently strong to push the debate in parliament to a rare *conscience vote* (a free vote that suspends the allegiance to one's political party that is characteristic of the Westminster system) and, even more rare, a vote that the prime minister and minister for health lost. Perhaps heartened by the success of this expression of dissent, others within government (and the Labor Opposition) subsequently began to challenge other aspects of the cabinet's conservative legislative agenda.

Following is an exploration of how and why such a tiny instance of biomedical technology could come closer to reactivating democracy than any other public expression of dissent has in recent years. What might the RU486 event reveal about the stuff of politics beneath the radar of much political theory, conscious intent, and public debate? The analysis involves examining how biotechnology might be located within two reformulations of the political: first, Michel Foucault's idea of political technologies of bodies combined with his and Giorgio Agamben's formulations of biopolitics; and second, what we might call deconstructive phenomenological formulations of the political developed from Hannah Arendt through to Jean-Luc Nancy and Jacques Derrida. Both these paradigms of the political provide insight into the relation between bodies, technology, and the political that helps to explain the emergence of biotechnologies such as RU486 as sites of domination and contest in social and political life. Biopolitical analysis can explain why RU486 became a site for the *governance and control of life* and of the Australian population in the first place. However, by adding consideration of time and temporality to its idea of the political, deconstructive phenomenology extends the political beyond the concept of governance and control of bodies. In ways examined subsequently, on this model of the political, what biotechnology and democratic politics share, if understood as "innovative" rather than instrumental, is the potential to temporalize being(s) (human and nonhuman) and thereby maintain living being(s) open to an undetermined future. This concept of the political can thereby explain how a biotechnology like RU486 can emerge as a site of *contestation* of government authority: understood as *innovative* rather than instrumental, the technological and the political, in concert with the unpredictability of biomaterial life, challenge appeals to biological and sociopolitical determinisms that preempt a future continuous with the past. Biotechnologies do this, most obviously, although not necessarily, by challenging assumptions of biological destiny; democratic politics challenges ideological (sociopolitical) determinism by contesting control measures aimed at preserving tradition for the sake of realizing an ideal future for a nation. Viewing RU486 as political technology and an event that expresses the possibility of bringing together matter, affects, and meanings in new ways reveals this shared feature of technology and politics and helps explain the surprising political power of RU486 peculiar to the Australian context.

However, neither paradigm gives sufficient attention to the way that the being(s) that are the site of biopolitical governance inhabit a *systematically* inequitable playing field, and so they cannot explain why *particular* biotechnologies may become a site of contestation of government authority. Some bodies (categorized in terms of race, sex, class, etc.) are expected to give away the time that they live to support the reproduction and maintenance of the biological life of the population so that others are free to dwell in the realm of potentiality. Biotechnologies and democratic politics also potentially redress such inequities. Conversely, biotechnologies can become the site of contestation of government authority when that authority perpetuates or enhances rather than alleviates these inequities. Hence the analysis below aims to argue that RU486 in particular succeeded in reawakening democracy in Australia because it exposed the reality of, and the possibility of redressing, one kind of inequitable sociopolitical distribution of lived time. Specifically, the RU486 event enfranchised women's time (labor time and maternal time) as a gift that, under a conservative regime of biopolitical governance, has been increasingly taken for granted without acknowledgment or recompense. Following are some points sketched toward that conclusion.

Political Technologies of Bodies and Biopower

How RU486 became a site of governance of the life of the Australian population without anyone really noticing is explained by Foucault's understanding of the relation between technology, the political, and bodies.

First, on his account, the *political is technological* by analogy insofar as the targets of both disciplinary power and biotechnologies are bodies *and* both combine empirical and calculated methodologies of intervention with technical knowledge of bodies. Just as biotechnologies intervene into bodies at the muscular, neurological, or molecular level to reorganize corporeal processes, disciplinary power operates at the micro level of the body's movements, spatiality, and temporal rhythms to realign the body's forces and powers (Foucault 1979, 136–38). It is this combination of knowledge with technique that allows Foucault to deem disciplinary techniques within regimes of the governance of bodies "*political* technologies of the body" (26). Moreover, unlike exercises in sovereign power, disciplinary political technologies operate with the same banality as technology in general, that is, without a single coordinating agent with necessarily sinister motives.

Second, and conversely, *biotechnologies are political* insofar as they are mobilized within these disciplinary regimes and so participate in the reproduction of normalized, productive, useful bodies and compliant subjectivities that are compatible with a neoliberal political economy. Technologies of the body are political insofar as they are embedded within what Heidegger has called an instrumental "way of thinking" (or a way of "enframing" biomaterial life; Heidegger 1977).[5]

Foucault explains how this instrumental regime of governance and its disciplinary techniques impacts the biomaterial life of bodies:

> Discipline increases the forces of the body (in economic terms of utility) and diminishes these same forces (in political terms of obedience). In short, it dissociates power from the body; on the one hand, it turns it into an "aptitude," a "capacity," which it seeks to increase; on the other hand, it reverses the course of the energy, the power that might result from it, and turns it into a relation of strict subjection. (Foucault 1979, 138)

But disciplinary power does not exhaust, or even best characterize, the political dimension of biotechnologies. Though technologies such as RU486 can be coopted or rendered problematic in the service of the disciplinary production of "useful" bodies with enhanced "capacities" and "aptitudes," they rarely aim at "obedience," the other political aspect of disciplinary techniques that Foucault notes; rather, discourses surrounding biotechnologies would suggest that they aim at the *enhancement of life for its own sake.* With respect to a genealogy of a body, whatever else, scientifically speaking, a biotechnology does (stopping the course of pain, expelling a zygote, or speeding up neurological events, metabolic rates, or whatever), to the extent that technology is innovative, it is more likely to disrupt the disciplined docile body, undo that dissociation of power and the body said to characterize discipline, and reopen the body's forces toward new directions or onto what Heidegger would call *potentiality* (Heidegger 1962, 184–85). Prior to his analyses of disciplinary power and biopower, and following Nietzsche, Foucault (1994, 376–78) had referred to this phenomenon of the disruption and realignment of corporeal forces as "emergence" or the "singularity of events": the eruption of forces from the "nonspace" of the interstices of corporeal and social struggles with an attendant transformation of meaning. Though Foucault does not put it this way, biotechnologies, along with anything or anyone that touches a human body (including mediums of sociopolitical meaning), participate in this reopening of forces through a *retemporalization of the body.* In any case, characteristic of emergence or the singularity of events is diversity of corporeal temporalities.

However, these technologically enhanced bodies do not escape regimes of governance. As technologies for the enhancement of biological life, biotechnologies enter the second political register, besides disciplinary power, that Foucault claims is characteristic of modern liberal democracies: *biopower.* Alongside the government of anatomical bodies through disciplinary power, biopower consists in "interventions and regulatory controls" that exercise the "power to *foster* life or *disallow* it" in the interests of maintaining the "biological existence of a population" (Foucault 1980, 137–39). Biopower does not aim at an individual body, rendering it compliant; rather, biopower targets biological processes and the "life" of a "species body." But emergence, innovation, and diversity of human biological existence are not the aims of this governance of life. On the contrary, biopower aims at curtailing "random events" and "achieving overall equilibrium" in a population, an equilibrium that reassures with the promise of protecting "the security of the whole from internal dangers" (Foucault 2003, 249). Biotechnologies that enhance the unpredictability of the emergence of corporeal events would present as one such danger. As Foucault later suggests, biopower is "totalizing." Or rather, the combination of disciplinary power and biopower, under a political rationality that attends to the health and welfare of the individual to ensure the health, stability, uniformity, and security of the whole (pastoral power), is an "individualizing and a totalizing" form of state power (Foucault 2002, 332). This multifaceted political technology of bodies makes subjects in two senses: the individual is subjugated, subject to relations of control and dependence, and she assumes an identity "by a conscience or self-knowledge" in terms of existing social norms (331).

Though Foucault teases out the workings of biopower by focusing on discourses of sexuality (in volume 1 of *The History of Sexuality*) and race and racism (in *Society Must Be Defended*), the analysis is especially suited to explaining the political investment in the regulation of reproductive technologies (including the former Australian government's investment in controlling the licensing of RU486) and the related connection between national security and political intervention into the home or private sphere. A further point about political technologies of bodies, borrowed in part from the phenomenological tradition, helps to explain the connection. Agamben (1998, 1–14) argues, following Foucault but with reference to Arendt, that what characterizes biopolitics is a collapse of the classical distinction between *zoe* and *bios*. Arendt describes this distinction in the following terms: *zoe* refers to human existence as biological life and its passive cyclic reproduction, which, conventionally, is excluded from politics and contained in the private realm, where it is governed by both biological destiny and force that for Aristotle characterizes the government of the household. *Bios* refers to the human being as a political subject caught up in "historical and biographical" time between birth and death—a lived, linear temporality of duration that presupposes a human sociopolitical world of speech and action from which *bios* emerges but through which it is also contested (Arendt 1998, 96–97). I will depart from the idea that the temporality of *bios* is linear and that *zoe* is cyclic and will return to this contestation of cyclic (natural) and linear (historical) time. The point for now is that for Agamben, insofar as modern politics involves an indistinction between *zoe* and *bios*, biological life *(zoe)* is the target of political power, and *zoe* is not just thereby included in the political as the "principal object of the projections and calculations of State power" but is included as *"bare life"* (human existence as that which can be killed; Agamben 1998, 8–9).

Pulling back a bit from the extremity of Agamben's characterization of biopolitics in terms of the inclusion of *zoe* in the *polis* as the bare life,[6] his general analysis of the zone of indistinction between *zoe* and *bios* has two consequences of relevance to the politics of reproductive and other biotechnologies. First, government of the population implies and depends on direct intervention into the home, into the sphere of the reproduction of the body politic, or to put it in biopolitical terms, increased government of the processes of human conception and reproduction is justified in terms of maintaining the uniformity, and security, of the population. Second, whereas the human body has a foot in both camps of *zoe* and *bios* (in both biological life and the corporeal potentiality that is a condition of sociopolitical agency), the spread of biopower, insofar as it aims at the homeostasis of human biological life, tends to reduce *bios*, if not to bare life, at least to *zoe;* biopower reduces the body of emergence open to potentiality (and thus political agency) to biological life determined by natural forces. This in turn justifies a kind of authoritative government that would determine the future security and uniformity of a population in terms of what the government considers to be the biological destiny of the nation. This is the logical extension of what Foucault describes as totalizing government or what Arendt calls "totalitarianism": achieving equilibrium and security by making subjects or "mankind itself the embodiment of law," where law is understood to flow inevitably from "Nature or History" and secures a future continuous with the past (Arendt 1994, 460–62). Or, as Agamben (1998, 10) puts it, insofar as "politics knows no value (and consequently no nonvalue) other than life," democracy tends toward "gradual convergence with totalitarian states."

The Political Technology of RU486

Such an account of biopolitics and political technologies of bodies does help to explain how the Australian government had, in the period 1996–2007, so easily mobilized 1950s "family values" and surveillance of bodies in the home as the basis of not just the health of a population but also national security in general. It is in this context of the spread of biopolitics and totalizing government that it had maintained its control over (and effective ban on) RU486. At stake in this accumulative spread and totalizing effects of biopolitics is democratic pluralism in general, women's agency (not all or exclusively but in relation to reproduction in particular), and with this, women's bodies in the realm of *bios*.

First, there are many indications of the biopolitical nexus in Australian politics between security, family values, and reproductive technologies. The 1996 legislation that transferred control of RU486 from medical–administrative (the TGA) to government authority was merely an early sign of the Liberal-National government's interest in regulating sexual reproduction and promoting traditional family values. There had also been a gradual dismantling of publicly funded child care since the mid-1990s, and as part of an explicit exercise in nation building, the government introduced a $5,000 "baby bonus"—a cash payment to mothers on the birth of each new child (as the treasurer puts it, he wants "one for the mother, one for the father, and one for the nation"). One of many indications that the home that reproduces the nation is assumed to be the heterosexual, middle-class family home was the prime minister's (unsuccessful) attempt in August 2000 to intervene into federal antidiscrimination legislation (Sex Discrimination Act 1984) to allow states to exclude lesbians and single women from accessing in vitro fertilization facilities.[7] More successful was the attorney general's Marriage Legislation Amendment Bill 2004, which removes any doubt as to what kind of family is to be reproduced here: "marriage means the union of a man and a woman to the exclusion of all others, voluntarily entered into for life. Certain unions are not marriages. A union solemnised in a foreign country between: (a) a man and another man; or (b) a woman and another woman; must not be recognised as a marriage in Australia."[8]

Second, federal government moves that linked the reproduction and maintenance of biological life in the home to wider matters of health, national identity, and security included a multi-million-dollar advertising campaign in July 2001 that, as Kane Race has analyzed, transferred responsibility for the war against illicit drugs from government to the home.[9] With the home, and heterosexual mothers in it, reemerging as the place from which the identity and security of the nation could be assured, it was merely a formality that the closing of national borders against refugees in August 2001 was justified by an implicit appeal to family values in the children overboard affair[10] (the Australian family was thereby marked as Anglo-Celtic and most likely Christian). At the same time, the prime minister launched his party's election campaign ahead of its third term of government with the (winning) slogan of conditional hospitality that assumed a uniformity of the Australian species body: "we decide who comes to this country, and the circumstances under which they come." This conditional hospitality is also a feature of the government asserting its control over RU486: "we" (i.e., the government as the arbiter of biological life) decide who is born and welcomed into this political life and under what circumstances. It is also not surprising, then—as silly as it sounds—that the government relaunched its domestic "war against terror" in February 2003 with a fridge magnet that designated the home as the first place from which one should be "alert, but not alarmed" about the dangers of strangers.[11] This theme of political hospitality and the link it establishes between home and nation, *zoe* and *bios,* is central to the deconstructive phenomenological reformulation of the political to which I will return.

A second theme, aside from conditional hospitality, throughout these legislative and public relations exercises in totalizing and biopolitical government, is the assumption that security rests on restoring the "natural" order of the biological life of the Australian species body and its reproduction in the home. This appeal to nature as the proper determinant of culture (and therefore of national identity) effectively justifies ideological–political determinism and totalizing government. Any appeals to a natural body or biological life as the proper determinant of culture, that is, sociopolitical appeals to a future determined by the assumed "natural progression" of biological life itself, in turn risks justifying ideological–political determinism on the basis of the assumption that (selective) elimination by law of technologies or practices that intervene into the natural order of life itself will restore human bodies (and the nation) to biological destiny. This is a problem for democracy not just because ideological–political determinism under autocratic government is democracy's most obvious adversary but also because appeals to a natural order of "life itself," especially with respect to those involved in the reproduction of "nature" in the home, reduce the *bios* characteristic of political agents to *zoe* governed by biological destiny.

In this context of government by appeal to the natural order, reproductive technologies present a particular problem and expose contradictions in the governance of biological life. On one hand, they directly challenge such government by raising the possibility of overturning nature and the biological through the technological. As Robyn Ferrell (2006, 33) argues, viewed as assisting "nature," reproductive technologies "instruct us in desires that are *impossible* in nature. In this way, reproductive technologies play their part in the political imaginary, and generally in 'biopower,' by cultivating the technological way of thinking in relation to reproduction, which has hitherto been its contrast." Though understood in terms of an instrumental link between the political (productive) and the biological (reproductive), between *bios* and *zoe*, reproductive technologies reveal the coconstitutive relation between these realms and hence the reproductive and political dimensions of both. Thereby, and on the other hand, reproductive technologies, viewed ontologically, challenge the assumption that the biological, in contrast to the political, is reproduced passively and uniformly by revealing the *innovative* and excessive dimension of both the biological and the political. Understood ontologically, and to adapt Heidegger's formula, reproductive technologies lie on a continuum of technologies that are innovative (rather than instrumental) "ways of thinking," where "ways of thinking," including political technologies, are not reducible to reflection and are *embedded in a chiasmic relation between meaning and being*, and so reorder the world that produces them.[12] In short, reproductive technologies confront political authority based on appeals to nature or *zoe* by challenging the distinctions between nature and culture, the domestic and the political realms, by exposing all reproduction (sexual and cultural) as both technological and political. Moreover, it "is only as a consequence of *political* technologies such as 'universal suffrage' and 'sexual equality' that these changes to reproduction can be conceived, let alone conceived of as desirable" (Ferrell 2006, 46). The question, then, that reproductive technologies raise, and the way they contest totalizing government, is which kind of political technologies should govern sexual and cultural reproduction: those that foster democratic pluralism and the diversity of emergent life events, or ministerial authority based on faith in the biological destiny of a nation?

This is the question that RU486 poses. In 1996 the main argument evoked in support of transferring responsibility for the licensing of RU486 from the TGA to the minister for health was that, as abortion is a social and moral issue rather than a purely technical or medical matter, any technique that frees up its availability requires the ongoing scrutiny of government. The opposite argument, but utilizing the same instrumental thinking (that the technomedical consists of an instrumental link between the sociopolitical and the biological), was applied in support of repealing the minister's authority in 2006—that RU486 is about providing women with a safe medical alternative to surgical termination, and not about the social impact or moral status of abortion per se. However, given that surgical abortion has been legal in Australia since the 1970s and has broad social acceptance, and given that the safety of the drug was, by 2005, no longer in serious question, at stake in the RU486 event of 2005–6 was neither the morality nor the health hazards of technical intervention into the biological. If reproductive (and other biomedical) technologies are always both ontological and sociopolitical, what the battle over RU486 is really about is women's agency, in two senses: first, in the obvious sense that from the perspective of government, retaining ministerial authority over (and the ban on) RU486, while not preventing women's reproductive choice per se, is an attempt to minimize the possibility that women could "do it themselves" away from direct scrutiny of biopolitical regulative mechanisms. RU486 is also about agency in a second, less obvious, sense: as a technology that simultaneously enacts a retemporalization of the body, a rearrangement of corporeal "powers," and a transformation of meanings to do with sex, conception, reproduction, and so on, RU486 maintains the body open to potentiality or the *bios*. Of course, pregnancy also effects a transformation of meaning and being. The salient issue for human existence as *bios* is that whatever path a body takes, agency rests on the condition that that path is not forced or predetermined by appeals to a future continuous with the past, either in terms of biology or conservative ideology. The possibility of the contestation and transformation of the meanings and arrangements that govern us is also the precondition of democracy. The question of who should have authority over the regulation of RU486 or sexual reproduction in general is a matter of individual conscience, where conscience, following Nietzsche and Arendt, is understood, not in terms of self-knowledge, but as the expression of agency, of the contestation of meanings that govern us.[13] The regulation of RU486 would be appropriately subject to a conscience vote in parliament, not because it is a moral issue, but because it is a political issue that raises the wider question of individual agency and the possibility of dissent within and outside government.

In tying the contestation of meaning and the transformation of corporeal being to both the impact of a biotechnology such as RU486 and the agency that is a precondition of democracy, I have taken the analysis into the territory of a different reformulation of the political—that of deconstructive phenomenology. While biopolitical analysis goes some way toward explaining how RU486 may have been both one of the first and then the most recent battleground for democracy in Australia in the past decade, it does not explain why *this* technology rather than any other political, social, or biotechnology mobilized resistance to, and contestation of, totalizing governance of life. This is in part due to a tendency in Foucauldian analysis to equate the political with regimes of "government"—understood as the management of the "possible field of action of others" (Foucault 2002, 341)—and to focus on the way effects of the centralization of the government of power relations in state institutions produce normalized and compliant subjectivities (341). Hence, though Foucault's models of disciplinary and biopower do allow for contestation of the kind of individualizing and totalizing government he describes, this is in terms of resistance to, or disruption of, the relatively stable "governmentalized" mechanisms that direct the conduct of others. His idea of "emergence" mentioned earlier, for example, is described as an eruption of "new forces" and diverse corporeal temporalities from a space of confrontation of the "endlessly repeated play of dominations" (377). Or resistance is said to emerge from either "bodies and pleasures" that somehow have escaped disciplinary mechanisms of government (Foucault 1980, 157) or the idea that wherever there are relations of power, there is also resistance or "antagonistic" "points of insubordination" and "means of escape" (Foucault 2002, 346). In any case, the emergence of new forces or the "singularity of events" that would contest the status quo is, on Foucault's account, a matter of both confrontation and accident. Though it would be right to say that contestation of biopolitical government by RU486 was not coordinated by anyone and so is accidental in that sense, that this particular biotechnology thwarted government authority was, I think, significant and not simply the fortuitous consequence of an arbitrary eruption in a play of forces.

Political Hospitality and Its Temporal and Gendered Dimensions

That RU486 *in particular* became a site of contestation of biopolitical power in Australia might be better explained in terms of the deconstructive phenomenological idea of the political as community devoted to the welcome, disclosure, or exposure of the new, alterity, or stranger. This second reformulation of the political is not inconsistent with Foucault's account of biopolitics, and aspects of it are implied in what I have said so far. It differs in offering a *political ontology* that does not equate the political with governmentality. While a Foucauldian understanding of the relation between the political, technologies, and bodies pits the emergence of corporeal forces and diverse temporalities against the imposition of a necessary continuity and equilibrium of biological life by biopolitical and disciplinary government, deconstructive phenomenology deems the temporality of *emergence, or potentiality, the raison d'état of the political itself,* insofar as the political is democratic. With this ontology, human existence as *bios* is always more than *zoe,* mere biological life or a diversity of corporeal forces. On one hand, *bios* is conditioned by the sedimented world of meaningful and material relations into which it is born; *bios* is thereby historical and develops a pattern of existence that carries forward into its future a collective history or tradition. On the other hand, the human being as *bios* has an impact on this world and transforms the conditions that condition it. Through encounters with the physical environment, biotechnologies, and other persons, the human being as *bios* breaks with, or transforms, the past. The crucial point raised by deconstructive phenomenology, though, is that this being open to potentiality or an undetermined future is dependent on a human world (or a space of the political) that *welcomes this impact of the new.* Hence various thinkers from Arendt through Jean-Luc Nancy and Jacques Derrida describe the political as the space of *hospitality:* the space of being-with-others that is the disclosure of alterity, uniqueness, or the "singularity of events." In this way, rather than focusing on the totalizing, controlling, and normalizing tendencies of liberal democracies, deconstructive phenomenology provides a normative model of the ethical and ontological preconditions for democracy and justice.

This ethical dimension of the political is as important as the ontological dimension for an analysis of the relation between bodies, biotechnologies, and democracy. What matters for the beings that make up a democratic polity is not so much whether bodies, with or without the assistance of explicit bio or other technologies, are disciplined to develop capacities and aptitudes that allow one to live within a sociopolitical economy (some habit and corporeal ritual is necessary for any kind of social life). Instead, what matters for democracy is that these bodies are valued equally in their uniqueness (as a who rather than a what, as Arendt puts it), that they remain open to possibilities for existence and close to decision-making processes with regard to their fate. Providing the space for this "singularity of events" to take place is the raison d'état of the political as much as the governmental management and regulation of the biological life of populations. This should also be the guiding principle of the political regulation of the consumption of biotechnologies that supplement the body, alter its temporal flows, and thereby enhance (or not) one's being-possible. As Isabelle Stengers and Oliver Ralet (1997) point out in their analysis of the ethics of the "war on drugs," what matters in the democratic regulation of the consumption of drugs, whether medicinal, recreational, and/or illicit, is the same as what matters to democratic pluralism, that people not be reduced to things *(zoe)* either through prescriptive moralism or a "purely technical" solution badly formulated.

Following are three points that elaborate how deconstructive phenomenology's model of the political might conceive of the relation between the political, technologies, and bodies. Understanding democracy as based on the principle of political hospitality provides the opportunity to unravel the link between and coconstitution of subjectivity, the home, the labor economy, the body as *bios*, and the nation. The point of this is to argue that the body, as the nexus of emergent corporeal forces and temporal flows that contest the status quo, is not outside of and in opposition to the regulation of time in public, social, and economic life; rather, it is as bodies *(bios)*, enabled through biomedical, social, and other political technologies, that human beings are opened to the realm of potentiality and are both threatened by and mobilized against biological determinism and authoritarian government. It is when government forgets the principle of hospitality in the regulation (or deregulation) of time that inequities arise, *bios* open to potentiality is stifled, and some bodies more than others suffer under the weight of attempting to live through multiple and conflicting time zones. The analysis provides a clue, for example, as to why patriarchal themes will tend to dominate in times of national insecurity, and why a failure of political hospitality will arise such that women as potential mothers will be targeted in the government of biological life. Inequities also arise in the partitioning and regulation of labor and leisure time. Considering *bios* open to potentiality to be dependent on political hospitality indicates that the ban on RU486 *combined with* the deregulation of the labor market, in the context of a failure of hospitality, puts sufficient pressure on lived time both to threaten and mobilize women's agency.

The first point to note is that in deconstructive phenomenology's view, the space of the political *provides the temporal conditions for political subjectivity*. Understood in terms of hospitality, deconstructive phenomenology presents us with an ontology in which the political provides the conditions that foster, not so much compliant subjectivity, but a subject open to potentiality and thereby capable of contesting the status quo. *Unconditional* hospitality, welcoming the *absolutely* (unknowable) other, uniqueness, or the "singularity of events" into one's home or one's political community, constitutes the home, the self, the nation, one's dwelling place as *open to the other*. This responsiveness, this welcome, this hospitality, *is* subjectivity; it *is* dwelling; it *is* the political. The temporal dimension of unconditional hospitality is this: hospitality is "desire" or "inclination" toward the uniqueness of others (Nancy 1991, 3–4) that institutes the "lapse" of time between a past tradition and an undetermined future (Derrida 2000, 127). It is this welcome of singularity or the "event," in other words, that contests totalizing government and its formation of compliant subjectivities discussed earlier. But, and this is the other side of the aporetic structure of hospitality, the welcome of the other or the "new" is not free from regimes of meaning embedded in sociopolitical institutions, including those of biopolitical and disciplinary government. Political and personal hospitality always carries *conditions* given by the *laws* of hospitality, by the ethos and interpretations of the culture and language the host has inherited and through which she assesses, knows, and welcomes the other. While neither the space of the political nor subjectivity can free itself from tradition to effect an absolute break between past and future, nevertheless the ethical imperative of hospitality is that in welcoming the other or the new, we must continue to experience its strangeness, its uniqueness, and allow it to interrupt and put into question our dwelling (Nancy 2002). In both cases of political and biotechnologies of hospitality, the "other takes place," the "singularity of events" emerges, through the interruption of lived, historical time and an accompanying transformation of the meanings and patterns of one's existence. This being-open to self-critique, prompted by expressions of the uniqueness, opens being to the undetermined future or potentiality. This is also, as Derrida suggests elsewhere, the condition of "democracy to come."[14] At least, this is the normative force and ethical condition of democracy on this model of the political.

Second, the political welcome of the "singularity of events" that disrupts the imposition of a future continuous with the past is not simply a matter of reflection, conscious dissent, or judgment. Central to this political subjectivity, inclined toward the uniqueness of others, is a *responsive body.* Arendt contributes much to the deconstructive phenomenological understanding of the political in *The Human Condition* through her account of the "web of human relations" or the "public sphere of appearance" conceived of in terms of the space of the prereflective disclosure of uniqueness through speech and action that maintains human beings open to potentiality (Arendt 1998). However, she tends to leave the temporalities of bodies out of account. She claims in *The Life of the Mind,* putting a Kantian slant on Nietzsche's idea of time, that it is critical thinking (judgment manifest as conscience) that "breaks up the unidirectional flow of time" between birth and death, opens a gap between the past as tradition, mediated by remembrance, and the future, mediated by anticipation, and propels us into infinite potentiality (Arendt 1978, 1:202–13). Since Arendt, others who have built on her account of the political insist that political hospitality is also a condition of the emergent, responsive body open to a world and that this body is a condition of the thinking, self-critical agent capable of contesting the status quo.

William Connolly, in his fine analysis of democracy and time, explains why it is important to consider the responsive body in calculations of the relation between the political, temporality, and technology (Connolly 2002). Whereas conservative political forces favor a slow pace of change, the continuous progression from past to future already discussed, the fast pace of modern life governed by a global economy and technologies that speed up the rate of activity can be equally conservative. Instead, a "certain asymmetry of pace," Connolly argues, "is critical to democratic pluralism" (143). In an alternative reading of Nietzsche's idea of time to that offered by Arendt, Connolly shows that this asymmetry of pace at all levels of life is lived by a body: accelerations in the pace of life disrupt habits, expose people to the "contingency and fluidity in cultural identity," and open us to ways of living that are "experimental," "more democratic and less fixed and hierarchical" (156). Connolly is aware of the dangers that such "uncertain experience of mobility in society" presents: those that resent such uncertainty may "press militantly to return political culture to a stonelike condition" (158). I believe this uncertainty describes the source of the conservative government of biological life that emerged in Australia in the decade leading up to the RU486 event. However, in articulating this point about the responsive body from the perspective of deconstructive phenomenology, I wish to stress the way that this body that lives "asymmetries of pace," and is thereby open to potentiality, is already caught within, and somewhat at the mercy of, the governmental partitioning and regulation of the pace of different zones of life. Particularly pertinent to this analysis is how, in the context of government that insists on a future continuous with the past and the severe conditional hospitality that this involves, different bodies negotiate the differing demands of the labor economy in relation to other demands of other time zones through which they are obliged to live.

Emmanuel Levinas's account of "The Dwelling" in *Totality and Infinity* elaborates the relationship between time, hospitality, and the body (Levinas 1969, 152–74). For Levinas, as with Derrida, the dwelling or home that conditional hospitality presupposes (from which one can welcome) is at once the *place* of one's dwelling (self, home, etc.) and the *event* and time of dwelling; that is, dwelling is the ongoing event (apparent in every activity, although exemplified for Levinas here by the activity of labor) that lifts the self, home, and so on, above an immediate affective relation to the world by punctuating time (or the "timeless"), thus introducing a difference between past and future. It is the body that labors, acts, touches, and is touched, that participates in and is crucial to the temporalization of time. For Levinas, the laboring body temporalizes space/place and time in two ways. The laboring body sets up a distance from the world on which it works while remaining grounded within the world that is not its fabrication. Hence there is an "equivocation of the body," the encumbered freedom of being "at home with oneself in something other than oneself," including in the world of meaningful relations into which we are thrown (Levinas 1969, 164). And second, the body that labors effects a "postponement" of the present and thus "opens the very dimension of time" (165). This "ambiguity of the body is *consciousness*" and "to be conscious is precisely to have time," a past to be remembered and a future anticipated (165–66).

The ambiguity of the body at the heart of the temporalization of place and time is the human being as *bios.* As Nancy points out, this body is already technological in the innovative sense of being open to *technē,* that is, to a community of bodies, regimes of meaning, prosthetic devices, biomedical supplements, and so on (Nancy 1993). As such the technological body effects the spatiotemporalization of being open to potentiality. It is not only always technological; *bios* is already political or dependent on the space of hospitality or the welcome. Hospitality, where I experience the other's strangeness or uniqueness, is how Nancy characterizes the impact of biotechnologies on one's body—his analysis of his experience of a heart transplant describes the way biotechnologies open the body to potentiality (Nancy 2002). But equally, hospitality operates in the other direction, where the body open to potentiality participates in the "sharing of singularity" or the welcome of uniqueness that, for Nancy and Arendt, characterize political community (Nancy 1991; Arendt 1998). Levinas puts this hospitality in ethical rather than political terms; that is, the precondition to this temporalization of the place of dwelling or the home of hospitality is the unconditional welcome of the alterity characteristic of the "ethical relation": to have a home, world, and so on, "I must know how to give what I possess," including my self-possession; "I must encounter the indiscreet face of the Other that calls me [and the concepts or laws of hospitality I embody] into question . . . by opening my home to him" (Levinas 1969, 171). It is the encounter with the absolute other, the new, or the stranger that, by signifying a "lapse of time" or alterity that cannot be memorialized or anticipated, gets going the temporalization of space and time and its disruption of presence. Or, to put this another way, the body open to potentiality is dependent on political hospitality.

The third point to be made, however, is that it is precisely this *bios* that is jeopardized by a failure of political hospitality or the severe conditional hospitality that is a feature of conservative and authoritarian government. Even in the best of times, not all bodies are welcomed as expressions of uniqueness in the same way. Insofar as *zoe* and *bios* are aspects of human life that are traditionally separated and hierarchialized in political philosophy (a hierarchy reaffirmed in conservative political practice even when, to follow Agamben, *zoe* in the "private sphere" is included in the public as the primary object of political power), the welcome of the new that contests tradition and maintains beings open to potentiality relies on some people giving time to sustain and reproduce life itself so that others can be welcomed as expressions of uniqueness. (Traditionally, it is "women, slaves, and barbarians" that are aligned with *zoe,* as Arendt [1998, 27] puts it). And those who are expected to give time in this way often do so without themselves being given hospitality by someone else. By accounting for the gendered dimension of the hospitality that is said to be central to democratic pluralism, I will attempt to demonstrate that political hospitality, as impossible as it is, gives and takes time, but that the more that hospitality becomes conditional under conservative political forces intent on securing the nation, the more the time that it takes is given by, or taken from, women *within* that polity. This is partly because of the way that national security is made increasingly dependent on the assumed stability provided by women giving time to others in the home. There is a second force at work here: the deregulation of labor time (which includes blurring the difference between leisure time, home time, and work time) with the effect that even those women who have been privileged enough to be paid to contest the concepts and "laws of hospitality" we have inherited have little time to give—either to others in the home or to contesting the meanings that govern us, which is the business of democracy.

It is significant I think that in December 2005, when those four female senators proposed the bill to repeal ministerial control over RU486, two other bills were rushed through the Australian Parliament (with only minor internal dissent quashed by the requirement to vote along party lines). One involved amendments to so-called antiterrorism legislation, including, not only detention without charge and orders controlling the movement of anyone merely suspected of "terrorist activities," but also antisedition provisions against those inciting "disaffection against the Government" (Anti-Terrorism Bill 2005). The second piece of legislation (misnamed "workplace choices") put in place a radical deregulation of the labor market consisting in the implementation of individual work contracts with little provision for upholding existing entitlements including national wage scales or workplace conditions (Workplace Relations Amendment [Work Choices] Act 2005). While the first directly jeopardized political agency (but paradoxically attracted little opposition across the political spectrum), the second did so indirectly in terms of its potential to impact the self-management of lived labor time. Of course, there is no obvious reason that these two forces (the shoring up of the family home as the foundation of national security and the deregulation of labor time) would have a more adverse effect on women than men. But insofar as our traditional laws of hospitality are patriarchal and both forces intersect across one person, that person is more likely to be a woman. My guess is that the three legislative issues of December 2005 (control of sexual reproduction, quashing of dissent, and the prospect of further deregulation of labor time) *together* is what provoked these women to move on RU486 (the issue of governmental control over which some political leverage would most likely be gained). All three issues, however, bear on the future of democratic pluralism insofar as they are about the relation between lived time, the "lapse of time" necessary for agency, and *bios* as the body open to potentiality.

Both Levinas and Arendt indicate conceptually how political hospitality depends on some people giving time to others without equivalent support. In his account of labor (and in his account of *eros*), Levinas admits to a patriarchal dimension to this supposedly unconditional ethical welcome of the other. Before and apart from the play of unconditional and conditional hospitality is an interim and arguably ultimate precondition—the hospitality provided by Woman in the home. In effect, the condition of unconditional hospitality is that women give time so that others have time for agency, consciousness, labor, and hospitality; that is, Levinas says explicitly that "feminine hospitality" is "the condition for recollection, the interiority of the Home, and inhabitation" or labor, but that this welcome is *not* the alterity that contests my self-possession and that is welcomed in the ethical relation (Levinas 1969, 157). But he does not say why. I will suggest two possible ontological reasons why Levinas insists that this "feminine" hospitality must be a "discreet, silent absence" rather than an "indiscreet" contestation (155). First, for dwelling to take hold, for the body to belong to a world, as ambiguous, uncanny, and open to potentiality as this belonging is, it cannot be entirely under erasure from the contestation of the new, of the "indiscreet" other of unconditional hospitality. Or, as feminist theorists have more critically put a similar point, a capitalist economy presupposes, without acknowledgment, that the ambiguous autonomy of the one who labors is dependent on some stability provided by women's (or someone's) care of the affective encumbered body in the home. Second, and related to the first point, while the welcome of unconditional hospitality cannot be reciprocated without annulling the "lapse of time" necessary to maintain being open to an undetermined future, the subject of that welcome must also be welcomed unconditionally by someone else if he is not to disappear entirely into the timeless present of immediate affectivity. That this "someone else" is Woman (rather than what Levinas usually refers to as the sexually neutral Third Party), at least when it comes to labor and eros, and that she is not herself explicitly given the security of hospitality by someone else, highlights the patriarchal basis of our tradition of hospitality.

In *The Human Condition*, Arendt, through her concept of *natality*, indicates why, even though those who give time in the home, as a precondition to political hospitality, do not have to be actual women (as Levinas also insists), this is usually the case. *Natality* refers to the way that the fact of birth, our own and that of others, rather than being-toward-death, signifies a new beginning, uniqueness, or what others call alterity or the "singularity of events."[15] As a second-order signification of uniqueness, a new beginning, and the who of the person, it is natality that is disclosed through speech and action in political community (Arendt 1998, 178–79). Natality, then, also refers to the way that the very appearance of the human being in a world temporalizes time by disrupting what Arendt considers to be the cyclic time of nature *(zoe)* and passive passing of linear historical and biographical (lived) time. And this space of the political that is the disclosure of natality is the "power as potentiality" where, in being together, *bios* is opened to an undetermined future rather than the materialization of the future in the present (201). However, this expression and preservation of natality, which is also the principle of democratic pluralism, also depends on women giving birth and time. Put simply: women give birth and consequentially, women, at least traditionally, are expected to give the time to others necessary for natality to appear and thereby temporalize, by disrupting, space and time. Arendt does not make any connection between giving birth and giving time. While women give birth and time in the first order, this seems to be forgotten in Arendt's (and most other) models of the political, such that the appearance of men and women in the public world through action and speech is usually described, including by Arendt herself, as a "second birth" (176). Apparently no one had to give birth or time for this second birth to happen. Both she and Levinas in different ways thus disavow any connection between giving birth and giving lived time, and they underplay the relation between the opening of the human being into potentiality and the embodied living of historical time. Insofar as these connections are overlooked, women of the *bios* are reduced to *zoe* through the expectation that in giving birth they will give lived time to others in the service of transforming, for others, the cyclic time of *zoe* into the supposed linear time of historical patterns of embodied existence that are then available for temporalization toward an undetermined future within political community.

RU486 Giving Back Time and Democracy

Lisa Guenther (2006) has gone some way in reestablishing the connection between giving birth and giving time in her diagnosis of the politics of reproduction. By highlighting the way *gestation* postpones the arrival of the future–present, she shows how the expectant mother gives (lived) time to allow the child to whom she gives birth to be an expression of natality. For the expectant mother, the child signifies natality in the sense of both the unknown future and the stranger—"a future that does not belong [to her], but for which [she is] nevertheless responsible" and which she welcomes into the home (Guenther 2006, 100). It is "gestation [that] marks immemorial . . . lapse of time" *and* the future welcome (99). Guenther also makes a connection between the opening of the human being into potentiality and the embodied living of historical time by remarking that "the pregnant woman already inhabits a world . . . [and] into this time of representation and consciousness, the anarchy of birth erupts" (100). Or, to put that point another way, the maternal body is the bearer of historical time, a lived temporality and mode of belonging to a world that is transformed, extended, and disrupted through gestation and birth. Without this maternal body giving lived time for "immemorial time" and an undetermined future, birth would not signify natality to anyone. Though Guenther does not make this point, what her analysis also suggests is that this reproductive body is not exclusively in a realm of *zoe* governed by biological destiny; it is an innovative reproductive technology in the realm of *bios.* To be acknowledged as such, however, the space of the political that maintains beings open to potentiality and an undetermined future must include extending reproductive choice to women. Hence Guenther concludes that even though the mother's gift of birth and gift of time is not chosen ("the gift of time does not originate in me . . . rather the giving of time *to* the other is made possible *by* the other" [102]), nevertheless, neither giving time nor giving birth should be forced (148, 150). As Barbara Baird argues, "withholding of the cultural and material means that enable certain performances" such as nonpregnancy, the "prohibition or even the limitation of abortion by law," significantly determines the pregnant woman's future self (Baird 2006, 123–25). Equally, only if potential mothers are also understood at expressions of natality (open to possibility) is it possible to hold open the space in which the political disclosure of natality is possible for everyone. So it was the bodies of women, the temporality of those bodies, and the time they give that was at stake in the battle over RU486. And it is because these bodies dwell in the realm of *bios* that they could also be a site of contestation of authoritarian biopolitical government.

Politics that excludes the first order of the welcome of natality from the benefits of the mutual disclosure of uniqueness between persons of equal value, which Arendt says characterizes the second order of political community, puts at risk the preconditions of democratic plurality. Though not addressing the gendered dimensions of the temporalization of time, Arendt does explain why and how conservative elements tend to close down political hospitality or at least make it severely conditional. Political conservatism kicks in when the "frailty of the human condition" (i.e., the "boundlessness" and "unpredictability" characteristic of the welcome and disclosure of natality whereby culture is contested, interrupted, and transformed) is felt as insecurity and uncertainty (Arendt 1998, 190–91). According to Arendt, this is most likely to happen when the public sphere is dominated by a science of process and predictability and when instrumental thinking and securing the future become the principles of government (Arendt 1998, 232). What I suggest is that under these conditions, responsibility for reigning in the unpredictable (i.e., expressions of natality that supposedly characterize our humanity) is siphoned off into the home. This perhaps explains why, in the worst of times, politics would eliminate felt instability, not only through war, border closures, censorship, and racism, but also through appeal to family values and the security that women are assumed to provide by giving time to others in the home. And second, insofar as the deregulation of labor time and the "busy time" of bureaucracy is accompanied by a simultaneous reassertion of the patriarchal themes of our tradition of hospitality, only some of us ("women slaves and barbarians") are expected to take responsibility for this time, thus freeing up time for others to partake in chance encounters with the "new" and in the contestation of traditional norms from which the new emerges.

Arendt's concern about the domination of the political by a "science of process and predictability" is with the way that the power of potentiality that characterizes the political space of democratic pluralism is reduced to force and the agent gives up her encumbered freedom under the assumption that her actions are determined

(Arendt 1998, 234). This is also a concern for us and for the women in the Australian government and why reclaiming their agency through repealing ministerial authority over RU486 amounted to also enfranchising women's time, the gift of which is increasingly taken for granted.

To better understand why particular technologies, bodies, and practices become sites of contest of biopolitical and authoritarian government, it is necessary to reassert a philosophy of the body between the biological determinism of classical materialism (that would conceive of bodies and reproduction in terms of mere life, or *zoe*) and idealism (including some models of "social constructionism") that would acknowledge that human bodies are in the realm of *bios*, conditioned by the sociopolitical environment into which we are thrown, but that would have difficulty finding a way through a political determinism that can arise from idealism and would, in foreclosing an unforeseeable future, reduce *bios* again to *zoe*. Such a philosophy of the body would acknowledge, along with Agamben, a "zone of indistinction" between *zoe* and *bios* but without reducing one to the other in the service of either biological or political determinism. This philosophy of the body would also acknowledge that the temporality of *bios* is the uneven and diverse temporality of lived rhythms and patterns of existence that are conditioned by the material and sociopolitical meanings into which we are thrown, but is not reducible to, or calculable in terms of, those conditions. Furthermore, this philosophy of the body acknowledges an irreducible link between this lived time of the zone of indistinction between *zoe* and *bios* and the interruption of lived time by the new, an interruption necessary for agency, that is, for the contestation of both political and biological determinism. Finally, it is necessary to stress that time can be given, possessed, spared, expanded, and diminished because time is lived by a body, and it is in the giving of lived time that justice and democracy take place (or not).

Being given time by others such that one has time to welcome the new should not be considered a privilege in a democracy. However, it is this feature of political hospitality and democratic pluralism that is under threat by a severe conditional hospitality that is currently in play. To counter these forces, the space of political hospitality that deconstructive phenomenology imagines, the space for the preservation of the expression of uniqueness, requires supplementation by at least two initiatives: first, this space for the contestation of the conservative concepts that govern us must be extended to all throughout the social fabric; and second, giving time to others necessary to give them time to be expressions of, and in turn to welcome, the new must be the responsibility of everyone, not just "women, slaves, and barbarians." RU486 is a biotechnology that has the power to facilitate such initiatives, not because it has the "power to foster or disallow life," in this case the life of a particular conglomeration of cells—understood in those terms, the decision to abdicate governmental authority over its licensing becomes merely a decision to transfer the work of eugenics from government to women; rather, RU486 has the power to spread democratic participation, because along with its impact on bodies and in concert with the democratic, feminist, and other political technologies that spawned it, RU486 has the power to disrupt the gift of time lived by a maternal body that would otherwise be rendered obligatory, or at least applauded as morally worthy, under a regime of authoritarian biopolitical government.

Notes

1 The senators involved were Lyn Allison (leader of the Democrats and senator representing the state of Victoria), Claire Moore (Labor Party, Queensland), Fiona Nash (National Party, New South Wales), and Judith Troeth (Liberal Party, Victoria). According to Nash, this is the first time in the history of the Australian Federal Parliament that four members of different political parties have cosponsored a private members bill in the Senate.

2 *Sydney Morning Herald*, February 16, 2006. RU486, or mifepristone, is a synthetic steroid that, in combination with a prostaglandin analog, provides what is widely considered to be a safe, do-it-yourself medical alternative to surgical abortion (De Costa 2005). It was developed in France in the 1980s and was licensed in France in 1988, the United Kingdom in 1991, and the United States in 2000, and it is available in many other countries such as Russia, China, Israel, and much of Western Europe.

3 The Australian government to which I am referring throughout the chapter is the Liberal-National (conservative) government that held power through four federal elections in 1995–2007.

4 After the Liberal-National government won a majority in both houses of parliament in 2004 (the first time any party has held absolute political power in Australia since the early 1980s), it tended to push through particularly conservative and controversial legislation without debate (e.g., "work-choices" legislation, which was opposed by the Labor Opposition and 90% of the Australian population). However, even prior to 2004, the government faced little dissent (including from the Labor Opposition) to some equally controversial legislation, despite widespread public opposition (e.g., the refugee legislation of 2001 and the decision to enter the war in Iraq in 2003).

5 For a detailed comparison of Foucault's and Heidegger's understandings of technology, see Rayner (2001).

6 Agamben's notion of "bare life" included in the polis in the "camp" or in the form of the "exception" (as that which can be killed) fits well with an analysis of how democratic states, under the sway of nationalism, harbor the potential for genocide and, to a certain extent, applies it to an analysis of Australia's policies involving the offshore processing and mandatory detention of refugees. But this schema works less well for an analysis of biopolitical government of the home and reproductive life. For an account of the limits of Agamben's biopolitics for the politics of reproduction, see Deutscher (2008).

7 For a summary of the legal issues raised by this proposed intervention, see Katrine Del Villar, "McBain v State of Victoria: Implications beyond IVF," Parliament of Australia Parliamentary Library, August 15, 2000, http://www.aph.gov.au/library/pubs/rn/2000-01/01RN04.htm.

8 Marriage Amendment Bill 2004, http://www.aph.gov.au/library/pubs/bd/2004-5/05bd005.htm.

9 The overarching slogan of the campaign was "the strongest defence against the drug problem is the family." The campaign included a series of TV advertisements set in the kitchens of multicultural, middle-class Australians depicting parents discussing the dangers of illicit drugs with their teenage children. Some advertisements were set on the streets depicting deaths of the same teenagers who had

not heeded the warnings. There was also a pamphlet distributed to every Australian household (titled "The Strongest Defence against the Drug Problem . . . Families") advising parents how to detect illicit drug use by their children and showing the importance of talking to them about the dangers of drug use. It is beyond the scope of this chapter to discuss the plethora of problematic assumptions underlying the campaign and possible reasons for its spectacular failure. For one such analysis, see Kane Race (2005).

10 The policy that closed Australian borders against asylum seekers was announced in August 2001 and targeted refugees (mainly from the Middle East) arriving by fishing boat from Indonesia. This involved the navy escorting these boats away from the coast of Australia, either back to Indonesia or to detention camps on islands offshore. To justify the policy and to shore up popular support, the minister for immigration and the prime minister announced, in October, that according to a report from the navy, people on one such boat were throwing their children overboard, presumably to prevent the navy from undertaking its mission. The announcement was accompanied by two visual images and comments suggesting that we do not want people with those sort of values to come to Australia. By the time it was revealed, several months later, that the report was false, it had had the effect that the government desired, and after several inquiries, no one has been held accountable for what has been put down to miscommunication. I have referred briefly to the "children overboard affair" in one analysis of the impact of this asylum-seeking policy on the fabric of Australian "community" (Diprose 2003). For a full analysis of this refugee policy and its intertwining with the "war on terrorism," see Marr and Wilkinson (2003).

11 This advertising campaign in February 2003 was centered on the slogans "let's look out for Australia" and "be alert, but not alarmed," and was similar in character to the antidrug campaign of 2000—a series of TV advertisements and a pamphlet (with fridge magnet) distributed to every Australian household advising members of the public to be alert to anything "unusual or suspicious in their neighborhood or workplace" and explaining what to do and who to call in the event that suspicions are raised. Again, the contradictions of the campaign, including problematic assumptions regarding what constitutes an "unusual or suspicious activity" and its connection to terrorism, begs further analysis. The counterstrategy from those

who thought the campaign ludicrous was to send the package back to the relevant government department and to incorporate the slogan "be alert, not alarmed" into everyday conversation in the form of a joke. My point in mentioning the campaign is to illustrate how comprehensive the former Australian government's conviction was that national borders can be secured and conditional hospitality controlled through the constitution of a particular kind of "home."

12 For analyses of reproductive technologies in support of this point, see, in particular, Ferrell (2006) and Franklin (1997).

13 Both Nietzsche and Arendt argue that "conscience" is not a manifestation of judgment that adheres to prevailing moral norms or laws but, on the contrary, is the expression of dissent—a challenge to and transformation (or revaluation) of the prevailing juridico-moral code. I have elaborated this idea of conscience and how its condition is a reflexive responsive body in Diprose (2008).

14 Derrida has written often about "democracy to come," but for his most recent clarification of the idea, see Jacques Derrida (2005, pt. I, chap. 8).

15 This concept of natality intervenes into an agenda set by Heidegger's (1962) *Being and Time* that would base singularity or alterity (innovation in meaning, thought, and being) in human finitude disclosed through being-toward death. For Arendt, this innovation or alterity that disrupts tradition arises, on the contrary, in being-toward-birth and the uniqueness this signifies.

References

Agamben, Giorgio. 1998. *Homo Sacer: Sovereign Power and Bare Life.* Trans. Daniel Heller-Roazen. Stanford, Calif.: Stanford University Press.

Arendt, Hannah. 1978. *Life of the Mind.* 2 vols. San Diego, Calif.: Harcourt.

———. 1994. *The Origins of Totalitarianism.* New Edition with added prefaces. San Diego, Calif.: Harcourt.

———. 1998. *The Human Condition.* 2nd ed. With an introduction by Margaret Canovan. Chicago: University of Chicago Press.

Baird, Barbara. 2006. "The Future of Abortion." In *Women Making Time: Contemporary Feminist Critique and Cultural Analysis*, ed. Elizabeth McMahon and Brigitta Olubas, 116–51. Crawley: University of Western Australia Press.

Connolly, William. 2002. "Democracy and Time." In *Neuropolitics: Thinking, Culture, Speed*, 140–73. Minneapolis: University of Minnesota Press.

De Costa, C. M. 2005. "Medical Abortion for Australian Women: It's Time." *Medical Journal of Australia* 183, no. 7: 378–80.

Derrida, Jacques. 2000. *Of Hospitality: Anne Dufourmantelle Invites Derrida to Respond.* Trans. Rachel Bowlby. Stanford, Calif.: Stanford University Press.

———. 2005. *Rogues: Two Essays on Reason.* Trans. Pascale-Anne Brault and Michael Naas. Stanford, Calif.: Stanford University Press.

Deutscher, Penelope. 2008. "The Inversion of Exceptionality: Foucault, Agamben, and 'Reproductive Rights.'" *South Atlantic Quarterly* 107, no. 1: 55–70.

Diprose, Rosalyn. 2003. "The Hand That Writes Community in Blood." *Cultural Studies Review* 9, no. 1: 34–50.

———. 2008. "Arendt and Nietzsche on Responsibility and Futurity." *Philosophy and Social Criticism* 34, no. 6: 617–42.

Ferrell, Robyn. 2006. *Copula: Sexual Technologies, Reproductive Technologies.* Albany: State University of New York Press.

Foucault, Michel. 1979. *Discipline and Punish: The Birth of a Prison.* Trans. Alan Sheridan. Harmondsworth, U.K.: Penguin.

———. 1980. *The History of Sexuality.* Vol. 1. Trans. Robert Hurley. New York: Vintage.

———. 1994. "Nietzsche, Genealogy, History." In *The Essential Works of Michel Foucault 1954–1984*, vol. 2, *Aesthetics, Method, and Epistemology*, ed. J. Faubion, 369–91. London: Penguin.

———. 2002. "The Subject and Power." Repr. in *The Essential Works of Michel Foucault 1954–1984*, vol. 3, *Power*, ed. J. Faubion, 326–49. London: Penguin.

———. 2003. *Society Must Be Defended: Lectures at the Collége de France 1975–76.* Trans. David Macey, eds. Mauro Bertani and Alessandro Fontana. New York: Picador.

Franklin, Sarah. 1997. *Embodied Progress: A Cultural Account of Assisted Conception.* London: Routledge.

Guenther, Lisa. 2006. *The Gift of the Other: Levinas and the Politics of Reproduction.* Albany: State University of New York Press.

Heidegger, Martin. 1962. *Being and Time.* Trans. J. Macquarrie and E. Robinson. New York: Harper and Row.

———. 1977. "The Question Concerning Technology." Trans. William

Lovitt. In *The Question Concerning Technology and Other Essays*, 3–35. New York: Harper and Row.
Levinas, Emanuel. 1969. *Totality and Infinity*. Trans. Alphonso Lingis. Pittsburgh, Penn.: Duquesne University Press.
Marr, David, and Marian Wilkinson. 2003. *Dark Victory*. Sydney: Allen and Unwin.
Nancy, Jean-Luc. 1991. *The Inoperative Community*. Trans. Peter Connor et al. Minneapolis: University of Minnesota Press.
———. 1993. Corpus. In *The Birth to Presence*, trans. Brian Holmes et al., 189–207. Stanford, Calif.: Stanford University Press.
———. 2002. L'Intrus. Trans. Susan Hanson. *New Centennial Review* 2, no. 3: 189–201.
Race, Kane. 2005. "Recreational States: Drugs and the Sovereignty of Consumption." *Culture Machine* 7. http://culturemachine.tees.ac.uk/frm_f1.htm.
Rayner, Timothy. 2001. "Biopower and Technology: Foucault and Heidegger's Way of Thinking." *Contretemps* 2: 142–56.
Stengers, Isabelle, and Oliver Ralet. 1997. "Drugs: Ethical Choice or Moral Consensus." In *Power and Invention: Situating Science*, trans. Paul Bains, 215–230. Minneapolis: University of Minnesota Press.

9 Infrastructure and Event: The Political Technology of Preparedness

ANDREW LAKOFF AND
STEPHEN J. COLLIER

As a number of analysts have argued, contemporary citizenship is simultaneously political and technical (see, e.g., Barry 1999; also contributions to Ong and Collier 2005). Thus, for example, access to material systems of circulation—such as water, electricity, communication, and transportation—is critical to participation in collective life. Indeed, demands for such access are often sources of political mobilization. This collective dependence on what we might call "vital systems" also fosters new forms of vulnerability. Threats to the operations of these life-supporting systems may come from a number of sources: natural disasters, terrorist attacks, technical malfunction, or novel pathogens. The prospect of such catastrophic threats now structures political intervention in a number of domains. Exemplary instances in which the failure to protect the functioning of such systems has caused major political fallout include the outbreak of mad cow disease and the European system of food supply, the attacks of September 11 and the system of air transportation, and Hurricane Katrina and systems of flood management. In this chapter, we describe the development of technical methods to identify and manage these threats to vital systems. The prevalence of these methods—and the common assumption of their necessity—suggests one answer to the question, how are political demands materialized today in programs of technical response? Through such methods, a range of significant "things" is internalized within political reason.

The chapter describes how critical infrastructure—and specifically the vulnerability of critical infrastructure—has become an object of knowledge for security experts in the United States. The production of such knowledge, we will suggest, is one part of a *political technology of preparedness* that addresses itself to a variety of possible threats. This political technology generates knowledge about infrastructural vulnerabilities through the imaginative enactment of a certain type of event. By the term *political technology,* we indicate a systematic relation of knowledge and intervention applied to a problem of collective life (Foucault 2001).[1] In this case, the political technology of preparedness responds to the governmental problem of planning for unpredictable but potentially catastrophic events. It works to integrate an array of material elements—ranging from switching stations to chemical plants to oil pipelines and network servers—into political organization.

Such political attention to the material underpinnings of collective life is not in itself new or surprising. Since the eighteenth century, experts have seen "the government of things" as one of the central tasks of state rationality (Foucault 2007). Thus current approaches in science and technology studies (STS) that draw attention to the salience of material artifacts to politics follow a long tradition of technocratic thought. From the vantage of critical analysis, what is important to specify is *how,* at a given moment, such technical artifacts as electricity networks are taken up as problems of collective existence: according to what rationality, and with what aim, do material things become political?

The chapter begins with a brief description of current critical infrastructure protection efforts in the United States. These efforts focus on mitigating perceived vulnerabilities to potentially disastrous events. It then turns to a key moment in which this relationship between infrastructure and event was developed—cold war civil defense. Here the chapter describes how the practice of "vulnerability mapping" worked as a way of generating knowledge about urban life in the shadow of nuclear attack. The chapter then follows the trajectory of imaginative enactment as a planning technique during the cold war and shows how this method of generating knowledge about vulnerability gradually extended to other types of threat. In closing, we suggest ways in which this story about recent developments in security expertise might be linked to broader discussions of the contemporary politics of technology.

Infrastructure and the Problem of Vulnerability

In a 2003 essay on "Infrastructure and Modernity," Paul Edwards posed the question of how to link detailed studies of the underpinnings of large-scale sociotechnical systems—which focus on issues such as the negotiation of standards and the problem of interoperability between systems—to questions raised in social theoretical discussions that emphasize the centrality of technological systems to modern life (Edwards 2003). He suggested that the differences between these two scales of analysis—one emphasizing the micropractices of technical experts in specific domains and the other making broad, general claims about modernity and technology—should, in principle, be reconcilable. This challenge is similar to the one posed by the editors of this volume, who have asked contributors to "draw questions of science and technology more fully into political theory, and to bring political theory to bear more consistently on our understanding of scientific practices and technological objects." In what follows we try to address these challenges by focusing on a specific technical domain, but one that responds to a broad political problem.

We focus on how experts in the management of risk have addressed the vulnerability of complex sociotechnical systems as a problem of collective security. This problem of system vulnerability is implicit in many current STS discussions of infrastructure. For example, Geof Bowker and Leigh Star (1999) emphasize that infrastructure is a fragile accomplishment and point to moments of breakdown as sites in which the work of infrastructure suddenly becomes visible. From a different vantage, the problem of system vulnerability is also central to social theories of risk, as in the work of Ulrich Beck (1999) on risk society. Beck argues that the very sociotechnical systems that were initially built to sustain human well-being as part of modern social welfare programs now generate new threats. Our dependence on these vital systems—energy, transportation, communication—is, for Beck, a source of vulnerability. His examples of threats that come from infrastructural dependence include ecological catastrophes such as Bhopal and Chernobyl, global financial crises, and mass casualty terrorist attacks. Such hazards, he argues, can cause global, irreparable damage, and their effects are of potentially unlimited temporal duration.

Our point in turning to Beck's argument here is neither to endorse nor to criticize its accuracy as a diagnosis of contemporary politics;[2] rather, it is to note a striking parallel between his diagnosis and that of a subset of contemporary security planners in the United States. For Beck, there is a broad class of contemporary threats—catastrophic risks—that outstrip statistical methods of management and control because their occurrence is unpredictable and their impact is unbounded. Moreover, he argues, it is our very reliance on modern technological systems that makes us especially vulnerable to these threats. Similarly, emergency planners in the United States—and increasingly elsewhere—now emphasize the dangers that are posed by catastrophic events, given our dependence on vital systems.

In what follows, we describe how these security experts have come to understand infrastructural dependence as an internal source of threat—and the techniques they have developed to mitigate this vulnerability. These expert practices work to make normally backgrounded aspects of infrastructure visible—not by observing its breakdown but by simulating its disruption. The claim of the chapter is not, then, that our polities are more vulnerable than they once were; rather, it is that system vulnerability has become a central problem structuring the way that technical artifacts are integrated into political calculation.

Critical Infrastructure Protection

Let us begin by describing current *critical infrastructure protection* (CIP) programs. CIP is a major aspect of homeland security strategy in the United States and has analogs in a number of European countries (see, e.g., Dunn 2005). Explicit governmental efforts to catalog critical infrastructure, assess its vulnerability, and mitigate threats to it began in 1996 with the Clinton administration's Presidential Commission on Critical Infrastructure Protection, which was formed in the wake of the Oklahoma City bombing and emerging concerns about the linkages created by information systems among technical infrastructures. U.S. security planners recognized that interoperability—the goal of much infrastructure development—was not only a boon to efficiency but also a potential source of danger; they argued that the interdependence of multiple infrastructures—information, communication, finance, energy—could lead to cascading and crippling failures.

After the attacks of September 11, CIP came to the center of homeland security strategy. The USA Patriot Act defined critical infrastructure as "systems and assets, whether physical or virtual, so vital to the United States that the incapacity or destruction of such systems and assets would have a debilitating impact on security, national economic security, national public health or safety, or any combination of those matters" (Department of Homeland Security [DHS] 2003, 6). In 2006 the Department of Homeland Security (DHS) released its long-delayed National Infrastructure Protection Program (NIPP), which contained an impressively long list of the sectors to be managed under the rubric of the nation's "critical infrastructures and key resources" (DHS 2006). These sectors included agriculture and food, the defense industrial base, energy, public health, banking and finance, drinking water and water treatment, chemical plants, dams, information technology, postal systems and shipping, transportation systems, and governmental facilities. This was the "stuff" that was to be made an explicit part of the new politics of security.

The NIPP contained three basic elements:

1. *Infrastructure inventory.* It sought to create a base of knowledge about the critical infrastructures of the United States in their complex interdependence by creating a "national infrastructure inventory." This inventory would gather "basic information on the relationships, dependencies, and interdependencies between various assets, systems, networks and functions" (DHS 2006, 31).
2. *Vulnerability assessment.* It called for the development of methods for analyzing risk that could guide resource allocation. These methods included vulnerability assessment—the identification of "intrinsic structural weaknesses, protective

measures, resiliency, and redundancies" in critical infrastructure (DHS 2006, 38).

3. *Coordination and federal assistance.* It defined the scope of federal intervention in CIP. The federal government was to play a coordinative role in the autonomous efforts of local governments and private sector actors and would distribute funds to state and local governments according to a "risk-based" formula to rationalize the expenditure of resources.

Thus the basic characteristics of critical infrastructure protection included, first, a concern with the critical systems on which modern society, economy, and polity depend; second, the identification of the vulnerabilities of these systems as matters of national security; and third, the development of security interventions whose aim is not to deter or defeat enemies but to mitigate system vulnerabilities.

Our goal here is not to evaluate whether such programs have in fact been successfully implemented (indeed, they have not) but rather to characterize their underlying logic. CIP is exemplary of a distinctive form of collective security, one that emphasizes protecting vital systems against potentially catastrophic threats. One of the key features of this form of security is that it seeks to manage the consequences of a variety of dangerous events—including terrorist attacks, natural disasters, and epidemics. It does this through the development and implementation of preparedness measures such as early warning systems, contingency planning, and scenario-based exercises.

As we have argued elsewhere, the genealogy of vital systems security can be traced to strategic bombing theory in interwar Europe, which focused on attacking the "vital, vulnerable" nodes of an enemy's industrial system (Collier and Lakoff 2007). This interest in the vulnerability of enemy systems was then internalized in U.S. programs for continental defense both before and during World War II. During the early cold war, U.S. civil defense planners were especially concerned to develop techniques to mitigate these vulnerabilities. Here it is useful to enter into some detail to see how the problem of system vulnerability has sparked the development of a novel security technology.

Vulnerability Mapping

The basic elements of this political technology of preparedness were developed during the early cold war, in response to the threat of a surprise attack by the Soviet Union. At this stage, preparedness meant massive military mobilization in peacetime to deter or respond to an anticipated enemy attack. The nation would have to be permanently ready for emergency, requiring ongoing crisis planning in economic, political, and military arenas. Civil defense was one aspect of such preparedness. U.S. civil defense plans were developed in response to the rise of novel forms of warfare in the mid-twentieth century: first, air attacks on major cities and industrial centers in World War II, and then intercontinental nuclear war. As World War II came to an end, U.S. military planners sought to ensure that the country did not demobilize after the war, as it had after World War I. They argued that the lack of a strong military had invited the surprise attack on Pearl Harbor. Now the Soviet Union presented a new existential threat. To meet it, the United States would have to remain in a state of permanent mobilization.

The U.S. Strategic Bombing Survey, conducted between 1944 and 1946, reported on the consequences of air attacks in England, Germany, and Japan and the effectiveness of these countries' civil defense measures. It recommended shelters and evacuation programs in the United States "to minimize the destructiveness of such attacks, and so organize the economic and administrative life of the Nation that no single or small group of attacks can paralyze the national organism" (Vale 1987, 58). The report pointed to the need to disperse key industries outside of dense urban areas and to ensure the continuity of government after attack. As Peter Galison has noted, the survey led military strategists to envision the United States in terms of its key weak points—to see the territory in terms of a set of targets whose destruction would hamper future war efforts (Galison 2001).

Faced with the threat of a surprise nuclear attack in the era of total war, military planners sought to develop a distributed system of preparedness that would enable civilian industrial production facilities to withstand an attack and support a viable counteroffensive (Collier and Lakoff 2007). Civil defense authorities in the 1950s created methods for spatially mapping domestic vulnerabilities to the threat of atomic warfare and then delegating preparedness activities to various agencies—from local government to individual families.

The prospect of a nuclear attack raised a number of interrelated questions for cold war national security planners: how would the enemy conceptualize U.S. national territory as a set of targets? What kinds of preparations were appropriate for meeting the threat of nuclear attack? And who should be responsible for organizing them? Civil defense authorities developed an elaborate set of planning practices. One of these was a procedure for mapping urban vulnerabilities. This procedure is significant in that it pointed toward the development of spatial knowledge about what would later be called critical infrastructure.

Vulnerability mapping generated a new form of knowledge about urban life. As opposed to statistical knowledge about the condition of the population, such as epidemiology or demography, this form of knowledge was not archival—it did not track the regular occurrence of predictable events over time; rather, vulnerability mapping produced knowledge about events—such as a surprise nuclear attack—whose probability could not be known but whose consequences could be catastrophic. Such knowledge involved not the calculation of probabilities but rather the imaginative enactment of events for which civil defense services would have to be prepared and the detailed analysis of how urban features would be affected by such events. In the process of evaluating vulnerability, planners made the material features of urban life an object of detailed political calculation.

Vulnerability mapping assembled a set of techniques for visualizing industrial facilities and population centers as targets of potential attack and developing appropriate response capabilities. This procedure not only meant identifying likely targets of attack; it also involved the imaginative enactment of attack to generate knowledge of which capabilities were needed to survive and fight back.

The 1950 document *United States Civil Defense* outlined the process of vulnerability mapping in schematic form (National Security Resources Board [NSRB] 1950). The starting point was the identification of "critical targets." To identify these targets meant developing a new way of understanding U.S. national space: through the reconstruction of the point of view of the enemy. Before the era of total war, knowing the mind-set of the enemy had been important mainly for planning theater operations. Now, the question was much broader: how did the enemy conceptualize U.S. territory as a set of targets?

United States Civil Defense assumed that the enemy would plan an attack based on the same principles of strategic bombing that were at the center of U.S. air-war doctrine. As the manual put it,

> The considerations which determine profitable targets are understood by potential enemies as well as our own planners. Such considerations include total population, density of population, concentration of important industries, location of communication and transportation centers, location of critical military facilities, and location of civil governments. (NSRB 1950, 8)

According to the civil defense plan, it was the job of each locality to determine its needs in preparing for attacks on critical targets within its jurisdiction. Planning at the local level was to be conducted through the imaginative enactment of a potential attack. Such an enactment would enable local civil defense planners to envision the probable impact of an attack, anticipate civil defense planning needs, and conduct exercises that would help identify weaknesses in their preparations.

United States Civil Defense provided a "hypothetical attack problem" as an example of how to identify civil defense needs. The hypothetical attack problem was a scenario consisting of an "attack narrative": it described two atomic detonations over an imaginary city *x:* one an air burst at twenty-four hundred feet and one an underwater burst (NSRB 1950, 117). The narrative then laid out the immediate impact of the attack: the water surge and lethal cloud of radioactive mist from the underwater burst; the explosive impact of the air burst and the flash fires that spread out up to a mile from ground zero; the casualties, including fourteen thousand to seventeen thousand from so-called mechanical injury (i.e., from the blast itself), seven thousand to eight thousand burn cases, and one thousand to three thousand radiation sickness cases from the air burst. The attack narrative also indicated the damage that would be inflicted on communications, transportation, utilities, and medical facilities.

All this information was intended to provide planners with knowledge of the exigencies for which they would have to prepare. "The hypothetical attack problem," argued *United States Civil Defense,* "should be realistic in order to bring out planning requirements in all segments of civil defense operations. The planners should accept the assumed effects, and analyze their needs accordingly" (NSRB 1950, 114). A city's civil defense needs could be determined as the difference between the envisioned impact of the bomb and its current response capabilities.

A series of technical manuals published by the Federal Civil Defense Authority gave local officials detailed instructions on how to make civil defense plans in a given city. For example, a 1953 manual titled *Civil Defense Urban Analysis,* guided planners in estimating how an atomic attack on a specific part of the city, at a specific time, would affect the structures and population of the city. The manual provided a detailed, systematic approach to mapping urban vulnerabilities. Knowledge of such vulnerabilities could then guide resources toward areas of greatest need.

This manual is of interest as a scheme for the development of a new knowledge of urban life as *tenuous*—in part because of its dependence on complex technological systems. Civil defense authorities saw that in the era of total war, the systems that had been developed to support modern urban life were now sources of vulnerability to enemy attack. Health facilities, systems of transportation and communication, and urban hygiene systems—whose construction had been essential to modern social welfare provision—were now understood in a new light, as possible targets and as necessary aspects of any emergency response. The material underpinnings of collective life were to be known and managed according to a certain political rationality.

The manual's introduction specified its aim and scope: "since the primary purpose of a civil defense urban analysis is to provide the tools for undertaking realistic civil defense planning, all pertinent aspects of the city must be considered" (Federal Civil Defense Administration [FCDA] 1953, 1). These pertinent aspects were to be considered in terms of their significance in the event of a nuclear attack. The relevant urban features to be analyzed were outlined in a lengthy table that constituted an impressive catalog of the elements of a city, including land use, building density, industrial plants, population distribution, police stations, the water distribution system, the electric power system, streets and highways, streetcars, port facilities, the telephone system, hospitals, zoos, penal institutions, underground openings (caves and mines), topography, and prevailing winds.

The table also indicated the "significance" of these features for civil defense planning. Thus knowledge about land use could help in estimating possible damage to various city functions. Industrial plants were significant as potential targets of sabotage or bombing and as important elements in police and fire-control planning. Water distribution systems were a potential target of sabotage and might be destroyed or disabled by a nuclear blast; they were also critical to fire control plans and were needed for emergency provision for attack victims and civil defense workers.

After identifying these features, planners were instructed to juxtapose them against one another on a series of operational maps. The goal of such maps was to determine which "pertinent" urban features would actually become important in the case of an attack and to present information that would be useful to specific urban services in formulating their civil defense plans. Once planners had assembled maps of significant urban features, the manual outlined a technical method for analyzing how these features would be affected by a nuclear attack. Given that the precise form of attack could not be known in advance, one needed a tool for modeling an attack's impact that was "sufficiently broad and flexible to meet all possible conditions" (FCDA 1953, 8).

To develop such a tool, the planner began by performing a "target analysis" to determine an enemy's assumed aiming point. The goal was to figure out what type of bomb a rational enemy would use to hit the city's main targets, and where the bomb would strike, to calculate the overall damage it would cause. To find the assumed aiming point, planners were to map both the area of industrial plant concentration and the distribution of the population. One then used a transparent acetate overlay with concentric circles indicating the level of bomb damage at different distances from ground zero. By placing the overlay on top of the map of facilities and population, the planner could estimate the point of attack that would cause maximum destruction. This was the assumed aiming point—which served as "a logical center for the pattern of civil defense ground organization of the community as a whole" (FCDA 1953, 10).

The next step was to estimate the damage a given sized bomb hitting a certain point would inflict. The technical manual focused this analysis on two key features of a city: first, facilities, such as industrial plants, public works, utilities, and services, and second, population. In each case, the point was not simply to measure the bomb's impact (the number of buildings destroyed, the number of individuals killed, injured, or made homeless) but the city's vulnerability—the relationship between blast impact and response capabilities.

In the case of structures, the factors determining damage were the size of the blast itself and possible damage from an ensuing firestorm. Physical damage from the blast was estimated by drawing concentric circles moving outward from ground zero, using information from a document that had been prepared by the Atomic Energy Commission and the U.S. Department of Defense (1950) called *The Effects of Atomic Weapons.* This document, based on data gathered in Hiroshima and Nagasaki, provided tables indicating blast damage from various bomb sizes at given distances from ground zero. Fire damage depended on such factors as building density, construction materials, precipitation, and wind velocity: here the key question was whether a blast would become a firestorm by spreading among neighboring buildings, which would obviously increase the structural damage considerably.

The manual directed planners to look at the destruction of facilities that would be important for response: "For example, one police station may house all of the police broadcasting equipment and one electric station may have the only available transformer which can change voltage from a distant source of electrical power to the voltage used for distribution through the city" (FCDA 1953, 53). The impact of an attack on the population, meanwhile, could be estimated as a function of the size and location of the bomb blast; the resident population versus daytime population in a given area (and therefore what time of day the bomb struck); and the condition of warning: was it a surprise attack or was the population on alert?

To map the city's probable number and distribution of casualties, the first step was to represent the distribution of the city's population in the city at the time of attack on a map based on estimates of daytime migration patterns. This was then paired with a table (provided by the federal authority) of the estimated percentage of fatalities and nonfatal injuries in a zone, given the size of the blast and the distance of the zone from ground zero. Using this table and an acetate overlay with concentric rings extending outward from ground zero, the planner would then "record the fatal casualties, nonfatal casualties and uninjured as calculated for each ring and for the various bomb sizes" (FCDA 1953, 36).

With this information, the planner could then generate isorithmic maps: city maps plotted with curving lines indicating the level of fatalities in a given subsector. These maps made it possible to visualize the distribution of casualties over the geography of the city. This tool for envisioning blast impact was a flexible instrument for assessing blast damage in generic terms at different points. Such maps brought urban populations into view as a spatially distributed set of casualty figures so that plans could be developed to provide relief in the wake of attack such as emergency medical and housing services.

The vulnerability mapping procedure thus provided a map of the physical damage of a likely blast, the casualties that resulted from it, and its impact on critical facilities. But more, it was characteristic of a way of coming to know national space—and the material features of that space—in terms of threat, vulnerability, and response capacity. This was not yet "critical infrastructure protection," but its basic logic was in place.

The Scenario

Over the course of the cold war, ambitious civil defense plans such as massive shelter systems were never fully implemented because of a lack of political will, skepticism about efficacy, and concern about their strategic implications. Nonetheless, a subset of security planners continued to attend to the problem of system vulnerability—still in relation to the threat of Soviet nuclear attack. Here it is illustrative to turn to the cold war trajectory of "imaginative enactment" as a knowledge production technique.

The practice of developing scenarios of nuclear attack to measure and improve current readiness was made famous by Herman Kahn of RAND in his 1962 book *On Thermonuclear War.* Kahn exemplified a new type of security expert distinctive to the period: not the military strategist or the civil defense planner but the "defense intellectual," a civilian with expertise in a technical domain—for example, mathematics, economics, or operations research—who applied this expertise to advise the government on nuclear strategy during the cold war.[3]

Kahn argued that for the strategy of deterrence to work, the enemy had to be convinced that the United States was prepared to engage in a full-scale nuclear war and had thus made concrete plans both for conducting such a war and for rebuilding in its aftermath. He criticized military planners for their failure to concretely envision how a nuclear war would unfold. If planners were serious about the strategy of deterrence, they had better be prepared to actually wage nuclear war. It was irresponsible not to think concretely about the consequences of such a war: what civil defense measures would lead to the loss of only fifty million rather than a hundred million lives? What would human life be like after a nuclear war? How could one plan for postwar reconstruction in a radiation-contaminated environment?

In the quest to be prepared for the eventuality of thermonuclear war, Kahn counseled, every possibility should be pursued. "With sufficient preparation," he wrote, "we actually will be able to survive and recuperate if deterrence fails" (Ghamari-Tabrizi 2005, 231). Kahn honed a method for what he called "thinking about the unthinkable" that would make such planning possible: scenario development. Like the civil defense attack narrative, Kahn's scenarios were not predictions or forecasts but opportunities for exercising an agile response capability. They trained leaders to deal with the unanticipated. "Imagination," Kahn wrote, "has always been one of the principal means for dealing in various ways with the future, and the scenario is simply one of the many devices useful in stimulating and disciplining the imagination" (Kahn 1962, 145).

Through the development of detailed attack scenarios, Kahn envisioned a range of postwar conditions whose scale of catastrophe was a function of prewar preparations, especially civil defense measures. These scenarios generated knowledge of infrastructural vulnerabilities and led Kahn to proposals for mitigating them. For example, a radioactive environment could hamper postwar reconstruction unless there was a way of determining individual levels of exposure. Thus he recommended giving out radioactivity dosimeters to the entire population in advance of war so that postwar survivors would be able to gauge their exposure levels and act accordingly.

All-Hazards Planning: Toward a Generic Technology

Let us now quickly summarize the process through which the methods of nuclear attack preparedness we have been describing became part of a more general political technology oriented toward multiple types of threat. Practices of civil defense were extended from nuclear attack to other types of disasters in the 1960s and 1970s through the advent of "all-hazards" planning. Beginning in the mid-1960s, state and local agencies—under the rubric of emergency management—sought to use federal civil defense resources to prepare for natural disasters such as hurricanes, floods, and earthquakes. Despite its different set of objects, the field of emergency management was structured by the underlying logic of civil defense: anticipatory mobilization for disaster. In the 1960s, state and local civil defense officials took up a number of the techniques associated with attack preparedness and applied them to natural disaster planning. These techniques included monitoring and alert systems, evacuation plans, training first responders, and holding drills to exercise the system.

Civil defense and emergency management shared a similar field of intervention—potential future catastrophes—which made their techniques transferable. Moreover, a complementary set of interests was at play in the migration of civil defense techniques to disaster planning. For local officials, federally funded civil defense programs presented an opportunity to support local response to natural disasters. From the federal vantage, given that civil defense against nuclear attack was politically unpopular, natural disaster planning developed capabilities that could also prove useful for attack preparedness. In the late 1960s, this dual-use strategy was officially endorsed at the federal level. Over the course of the 1970s, the forms of disaster to be addressed through emergency planning expanded to include environmental catastrophes, such as Love Canal and Three Mile Island, and humanitarian emergencies, such as the Cuban refugee crisis.

When the Federal Emergency Management Agency (FEMA) was founded in 1979, it consolidated federal emergency management and civil defense functions under the rubric of all-hazards planning. All-hazards planning assumed that, for the purposes of emergency preparedness, many kinds of catastrophes could be treated in the same way: earthquakes, floods, major industrial accidents, and enemy attacks were brought into the same operational space, given certain common characteristics. Needs such as early warning, the coordination of response by multiple agencies, public communication to assuage panic, and the efficient implementation of recovery processes were shared across these various sorts of disasters. Thus all-hazards planning focused not on assessing specific threats but on building capabilities that could function across multiple threat domains.

To operationalize all-hazards planning in the post-9/11 world, the DHS developed the National Incident Management System (NIMS). This is a system for deciding when a given event (an "incident of national significance," or INS) should trigger a temporary recomposition of governmental structures—and for governing how these temporary structures should operate. Multiple types of events can trigger the system: as the NIMS states, "For the purposes of this document, incidents can include acts of terrorism, wildland and urban fires, floods, hazardous materials spills, nuclear accidents, aircraft accidents, earthquakes, hurricanes, tornadoes, typhoons, war-related disasters, etc." This final *et cetera* is worth emphasizing: it indicates the expansiveness of the category of the INS.

Contemporary Preparedness

The political technology of preparedness thus addresses a variety of events that threaten vital systems, including natural disasters, terrorist attacks, epidemics, and technological accidents. What these potential events have in common is that they are considered low-probability, high-consequence threats. It is not possible to gather knowledge about them based on archival records of their occurrence; nor can they necessarily be deterred or prevented. Security interventions must then anticipate their occurrence.

Here imaginative enactment as a way to generate knowledge about current needs in the face of future events remains a central tool. As an example, we can look at a 2004 Homeland Security Council document called *Planning Scenarios*. This was a set of fifteen disaster scenarios to be used by DHS as "the foundation for a risk-based approach" to homeland security planning. These possible events—including an anthrax attack, a flu pandemic, a nuclear detonation, and a major earthquake—were chosen on the basis of plausibility and catastrophic scale.

The scenarios were not predictions or forecasts; rather, they made it possible to generate knowledge of current vulnerabilities and the capabilities needed to mitigate them. As one expert commented, "we have a great sense of vulnerability, but no sense of what it takes to be prepared. These scenarios provide us with an opportunity to address that" (David Heyman, director of the homeland security program at the Center for Strategic and International Studies, quoted in Lipton 2005). Using the scenarios, DHS developed a menu of the "critical tasks" that would have to be performed in various kinds of major events; these tasks, in turn, were to be assigned to specific governmental and nongovernmental agencies.

Scenarios and scenario-based exercises are widely used in governmental and para-governmental preparedness efforts in the United States and elsewhere. They involve enactments of varying detail and scale, followed by reports on the performance of response. They are often designed by policy institutes and think tanks under contract to government agencies. In 2001, "Dark Winter" was performed, a scenario depicting a covert smallpox attack in the United States. This was an "executive-level simulation" set in the National Security Council over fourteen days. Current and former public officials played the roles of members of the National Security Council, and members of the executive and legislative branches were briefed on the results. One outcome was the Bush administration's decision to produce three hundred million doses of smallpox vaccine.

"Silent Vector" (2002) was an exercise in how to deal with the threat of an impending terrorist attack when there is not enough information to provide protection against the attack. The president, played by former senator Sam Nunn, was told of credible intelligence indicating an upcoming attack on the nation's energy infrastructure but was not given any information on where or when the attack would take place. Other examples include 2003's simulated anthrax attack, "Scarlet Cloud," "Black Dawn," which simulated a prospective al-Qaeda nuclear attack, held in Brussels in 2004, and the biennial TOPOFF exercises held by the DHS. TOPOFF 3 was enacted in April 2005 and included a car bombing, a chemical attack, and the release of an undisclosed biological agent in New Jersey and Connecticut. It was the largest terrorism drill ever, costing $16 million and including ten thousand participants. The event also included a simulated news organization, which was fully briefed on events as they unfolded.

In the January 2005 "Atlantic Storm," former secretary of state Madeleine Albright played the U.S. president in an exercise simulating a smallpox attack on multiple nations of the transatlantic community.[4] Istanbul, Frankfurt, Rotterdam, and multiple U.S. cities were hit. In a mock summit, former prime ministers of European countries played the role of heads of state. Questions of immediate response were posed: what kind of vaccination approach to use? Which countries have enough supplies of vaccine, and will they share them? Will quarantine be necessary? After the exercise, participants concluded that, first, there was insufficient awareness of the possibility and consequences of a bioterrorist attack; and second, no organization or structure is currently agile enough to respond to the challenges posed by such an attack. Structures of coordination and communication of response in real time must be put into place. The exercise produced a sense of vulnerability to new threats among participants.[5]

The conclusions were similar to those of other such exercises: governments are not adequately aware of or prepared for catastrophic events. Secretary of Homeland Security Michael Chertoff said of TOPOFF 3, "We expect failure because we are actually going to be seeking to push to failure" (DHS 2005). In producing system failure, scenario-based simulations generate knowledge of gaps in needed capability. These can then be the target of intervention. In so doing, they forge new links—communicational, informational—among various agencies: local and national government, public health, law enforcement, intelligence. These exercises are part of an effort to develop an integrated system for assigning priorities and allocating resources in preparation for emergency.

Thus the practice of linking possible future events to current vulnerabilities in vital systems is now widespread. Indeed, it is possible to map a growing field of "preparedness expertise" that develops this knowledge and makes recommendations for intervention. One might look, for example, at the work of the port security expert Stephen Flynn (2005), the public health expert Irwin Redlener (2006), or the natural disaster specialist Lee Clarke (2005)—or at reports produced by places like the RAND Center for Terrorism Risk Management Policy, for instance, the recent *Considering the Effects of a Catastrophic Terrorist Attack*. This report is based on a scenario in which terrorists conceal a ten-kiloton nuclear bomb in a shipping container and ship it to the Port of Long Beach, where the bomb explodes. While the report describes the massive death and destruction such a bomb might cause, it is mainly focused on the economic impact of this disruption of the global shipping supply chain. The report does not predict such an attack or calculate its likelihood. As the authors write, "We used this scenario because analysts consider it feasible, it is highly likely to have a catastrophic effect, and the target is both a key part of the US economic infrastructure and a critical global shipping center" (Meade and Molander 2006, xv).

We can make several points about this type of enactment. First, it is not a prediction, forecast, or model of how the future will unfold but rather an intervention in the present. Second, its purpose is not to provoke public anxiety or militate toward an intensified war on terror; rather, it is to generate expert knowledge about what the event would entail: as the RAND report states, it enables policy makers "to anticipate the types of decisions they might be called upon to make, reflect in time of relative calm on their options, and plan well in advance for contingencies" (Meade and Molander 2006, xviii). And third, it does not apply only to terrorism; the technique is also brought to bear to approach dangers such as avian flu, earthquakes and hurricanes, and environmental catastrophe. For example, the 2006 "Strong Angel" scenario exercise in San Diego combined an avian flu pandemic with a cyberattack, focusing on generating knowledge about how to design information systems for use by military and civilian organizations in humanitarian emergencies (Markoff

2006). And as is well known, FEMA had contracted a private firm to develop hurricane scenarios on the Gulf Coast prior to Hurricane Katrina—though DHS cut the program's budget so that the exercises were never conducted.

This leads us to a final point: failures of response do not undermine the norm of preparedness but rather intensify it—as we could see after Katrina, in political demands for better preparedness. This is characteristic of a political technology: it defines and regulates targets of intervention according to a normative rationality (see Rabinow 2003). In this case, the imagined enactment of events of a certain type—low probability, high consequence—makes it possible both to generate knowledge about vulnerabilities and develop techniques for mitigating them.

Conclusion

Techniques for generating infrastructural knowledge that were initially assembled as part of civil defense are now applied to planning for various types of disaster—hurricanes, floods, terrorist attacks, epidemics—and not only by the U.S. government. In conclusion, let us turn to the question of how attention to the ways that infrastructure is understood and managed by experts can be related to broader issues in the social theory of modernity.

Here it is useful to return to our earlier discussion of the work of Ulrich Beck on catastrophic risks. Beck argues that today, the very industrial and technical developments that were initially put in the service of guaranteeing human welfare now generate new threats. Our very dependence on critical infrastructures—systems of transportation, communications, energy, and so on—has become a source of vulnerability. For Beck, the danger emanating from technical developments such as nuclear accidents and genetically modified food shapes a more general perception that "uncontrollable risk is now irredeemable and deeply engineered into all the processes that sustain life in advanced societies" (Beck 2002, 39–56). Such dangers "abolish old pillars of risk calculus," outstripping our ability to calculate their probability or to insure ourselves against them.

As Francois Ewald points out, the precautionary principle has been an influential response to these novel forms of threat in Europe, especially those linked to the environment. In the context of possible catastrophe, Ewald notes, statistical calculation is no longer relevant—one must take into account not what is probable or improbable but what is most feared: "I must, out of precaution, imagine the worst possible" (Ewald 2002, 286). Thus a principle of precaution in the face of an incalculable threat enjoins against risk taking—for example, the implementation of new and uncertain technologies such as genetically modified food. In this manner, it seeks to keep the dangerous event from occurring.

In contrast, as we have described, a very different approach to uncertain but potentially catastrophic threats has emerged and extended its reach first in the United States and increasingly transnationally. Like precaution, it is applicable to events whose regular occurrence cannot be mapped through archival knowledge and whose probability therefore cannot be calculated. In contrast to precaution, however, this approach does not prescribe avoidance; rather, it enacts a vision of the dystopian future to develop a set of operational criteria for response. Preparedness does not seek to prevent the occurrence of a disastrous event but rather assumes that the event will happen. Instead of seeking to constrain action in the face of uncertainty, it turns potentially catastrophic threats into vulnerabilities to be mitigated. The technology of preparedness, as exemplified in programs such as critical infrastructure protection, thus brings a heterogeneous set of things into political reason. It is through this technology that experts and officials have come to see collective life as dependent on the functioning of a series of interdependent, complex, and above all highly vulnerable systems.

Notes

1 This use of the term *political technology* follows the work of Michel Foucault, who showed that technical practices for managing life have been central to politics since the late eighteenth century, with the advent of biopolitics. Modern polities, he argued, are integrated not through a community of shared values along the model of the Greek *polis* but rather through a "political technology of individuals" (Foucault 1998).

2 For critiques of the empirical validity of his claim that contemporary technological risks outstrip private insurability, see Bougen (2003) and Ericson and Doyle (2004).
3 For Kahn's biography, see Ghamari-Tabrizi (2005).
4 For a summary, see Smith et al. (2005).
5 As a German official said, "For someone who has been around in the security and defense fields in its traditional sense for many years, this was quite a surprising and breathtaking exercise. . . . This is something I think a very small minority of politicians in Europe are aware of." See http://www.atlantic-storm.org/.

References

Barry, Andrew. 1999. *Political Machines: Governing a Technological Society.* London: Continuum.

Beck, Ulrich. 1999. *World Risk Society.* London: Wiley-Blackwell.

———. 2002. "The Terrorist Threat: World Risk Society Revisited." *Theory, Culture, and Society* 19, no. 4: 39–56.

Bougen, Phillip D. 2003. "Catastrophe Risk." *Economy and Society* 32, no. 2: 253–74.

Bowker, Geoffrey, and Susan Star. 1999. *Sorting Things Out: Classification and Its Consequences.* Cambridge, Mass.: MIT Press.

Clarke, Lee. 2005. *Worst-Cases: Terror and Catastrophe in the Popular Imagination.* Chicago: University of Chicago Press.

Collier, Stephen, and Andrew Lakoff. 2007. "Distributed Preparedness: The Spatial Logic of Domestic Security in the United States." *Environment and Planning D* 25: 7–28.

Department of Homeland Security. 2003. *The National Strategy for the Physical Protection of Critical Infrastructures and Key Assets.* Washington, D.C.: Department of Homeland Security.

———. 2005. "Transcript of Press Conference with Secretary of Homeland Security Michael Chertoff on the TOPOFF 3 Exercise." http://www.dhs.gov/xnews/releases/press_release_0650.shtm.

———. 2006. *National Infrastructure Protection Program.* Washington, D.C.: Department of Homeland Security.

Dunn, Myriam. 2005. "The Socio-political Dimensions of Critical Information Infrastructure Protection (CIIP)." *International Journal of Critical Infrastructures* 1, nos. 2–3: 258–268.

Edwards, Paul N. 2003. "Infrastructure and Modernity: Force, Time, and Social Organisation in the History of Sociotechnical Systems." In *Modernity and Technology*, ed. Thomas Misa, Philip Brey, and Andrew Feenberg, 185–225. Cambridge, Mass.: MIT Press.

Ericson, Richard V., and Aaron Doyle. 2004. "Catastrophe Risk, Insurance, and Terrorism." *Economy and Society* 33, no. 2: 135–73.

Ewald, Francois. 2002. "The Return of Descartes's Malicious Demon: An Outline of the Philosophy of Precaution." In *Embracing Risk: The Changing Culture of Insurance and Responsibility*, ed. Tom Baker and Jonathan Simon, 273–301. Chicago: University of Chicago Press.

Federal Civil Defense Administration. 1953. *Civil Defense Urban Analysis.* Washington, D.C.: Federal Civil Defense Administration.

Flynn, Stephen E. 2005. *America the Vulnerable: How Government Is Failing to Protect Us from Terrorism.* New York: HarperCollins.

Foucault, Michel. 2001. "The Political Technology of Individuals." In *The Essential Foucault,* vol. 3, ed. James Faubion, 403–417. New York: New Press.

———. 2007. *Security, Territory, Population: Lectures at the College de France.* New York: Palgrave.

Galison, Peter. 2001. "War against the Center." *Grey Room* 4: 6–33.

Ghamari-Tabrizi, Sharon. 2005. *The Worlds of Herman Kahn: The Intuitive Arts of Thermonuclear War.* Cambridge, Mass.: Harvard University Press.

Kahn, Herman. 1962. *Thinking About the Unthinkable.* New York: Horizon Press.

Lipton, Eric. 2005. "U.S. Report Lists Possibilities for Terrorist Attacks and Likely Toll." *New York Times*, March 16.

Markoff, John. 2006. "This Is Only a Drill: In California, Testing Technology in a Disaster Response." *New York Times,* August 28.

Meade, Charles, and Roger C. Molander. 2006. *Considering the Effects of a Catastrophic Terrorist Attack.* Washington, D.C.: RAND Center for Terrorism Risk Management Policy.

National Security Resources Board. 1950. *United States Civil Defense.* Washington, D.C.: National Security Resources Board.

Ong, Aiwa, and Stephen J. Collier. 2005. *Global Assemblages: Technology, Politics and Ethics as Anthropological Problems.* London: Blackwell.

Rabinow, Paul. 2003. *Anthropos Today: Forms of Modern Equipment.* Princeton, N.J.: Princeton University Press.

Redlener, Irwin. 2006. *Americans at Risk: Why We Are Not Prepared for*

Megadisasters and What We Can Do. New York: Random House.

Smith, Bradley T., Thomas V. Inglesby, Esther Brimmer, Luciana Borio, Crystal Franco, Gigi Kwik Gronvall, Bradley Kramer, Beth Maldin, Jennifer B. Nuzzo, Ari Schuler, Scott Stern, Donald A. Henderson, Randall J. Larsen, Daniel S. Hamilton, and Tara O'Toole. 2005. "Navigating the Storm: Report and Recommendations from the Atlantic Storm Exercise." *Biosecurity and Bioterrorism: Biodefense Strategy, Practice, and Science* 3, no. 3: 256–67.

U.S. Department of Defense. 1950. *The Effects of Atomic Weapons.* Los Alamos, N.M.: U.S. Atomic Energy Commission, Los Alamos Scientific Laboratory.

Vale, Lawrence J. 1987. *The Limits of Civil Defense in the U.S.A, Switzerland, Britain and the Soviet Union: The Evolution of Policies since 1945.* New York: St. Martin's.

10 *Faitiche*-izing the People: What Representative Democracy Might Learn from Science Studies

LISA DISCH

DEMOCRATIC POLITICAL THEORY has unheralded champions in science studies. A cadre of scholars has succeeded in calling attention to a poignant paradox: modern democracy, which came to be in and through the fragile but ingenious practice of representative government, was rendered "powerless as soon as it was invented, because of the counterinvention of Science" (Latour 2004, 71). The idea is that the power of the laboratory was invented at the same time as that of the Assembly, and on the basis of a shared fiction of ontological difference. That fiction cleaves nature, as the domain of "mute things," from society, as the domain of "speaking humans," and divides the problem of representation in two (68).

This cleavage has four pernicious effects. It inaugurates (1) a qualitative distinction between the laws of necessity and those of ethics, (2) an epistemological difference and hierarchy between "the truth of things and the will of humans" so that it seems easier to say what entities *are* than what they *want* (Latour 2004, 148), and (3) a disciplinary divide between Science, which is reputed to have the power to specify what entities are, and politics, which is left groping for ways to say what they want. This difference between being and wanting deals the final blow because it sets up a contest that politics cannot win. This is its fourth pernicious effect: it pits the laboratory, as the domain of "indisputable matters of fact," against the Assembly as that of "endless discussion" (223). The result is what Bruno Latour has eloquently denounced as "the strange politics by which facts have been made at once completely mute and so talkative that, as the saying goes, 'they speak for themselves'—thus providing the great political advantage of shutting down human babble with a voice from nowhere that renders political speech forever empty" (Latour 1999, 140).

Political theorists have collaborated with this strange politics insofar as we have taken it for granted that representation poses a very different kind of problem in the Assembly than it does in the laboratory. Because we accord the citizen-subject a monopoly on speaking, we imagine political representation to be uniquely fraught by the complexities and uncertainties of discussion, whereas Science merely witnesses "mute things" morph into "speaking facts" by the alchemical agency of the experiment (Latour 2004, 68). Our conviction about this has produced in us an almost schizophrenic ambivalence toward scientific expertise. We begin by putting an overweening trust in nature to "adjudicate between claims and facts" (Braun 2002, 223). Then we entrust Science with the power to discover nature and so to settle disputes of policy and principle by the "unforced force" of natural necessity. When the sciences reveal that they cannot perform this "God trick" of nonpartisan vision, we castigate them for passing off their own artifacts as real. Science takes its place alongside all the other fetishes that we moderns have smashed with the busy jackhammer of skepticism.

Isabelle Stengers and Bruno Latour attempt to rewrite this story from its beginning in this bifurcation of the problem of representation into knowing the "truth of things" and debating the "will of humans" (Latour 2004, 148). They propose to view science and politics as working in a single "assembly of beings *capable of speaking*" and to define the "lab coats"—akin to politicians—as "the spokespersons of the nonhumans" (64; cf. Stengers 2000, 87). Scientists and politicians neither represent in the same way nor do the same kind of work. But they both confront the problem posed by what Latour calls the "ancient" political sense of parliamentary democracy where a constituency that is an agency elects another agency to speak on its behalf (Latour 2004, 41). As Stengers (2000, 61) puts it, "the same question presents itself with regard to the person who claims to speak for others as it does with regard to the theory that claims to represent the facts: 'How does one recognize the legitimate claimant?'"[1]

A legitimate claimant can emerge only from a "staging" or election that is free and fair, from a system of representation that respects the agency of both constituent and representative. The impetus behind importing the paradigm of political representation into the laboratory is to counter the governmentalist fiction of facts that speak for themselves and experts who transmit them. Stengers and Latour together counter the fantasy of the impartial expert with a more pragmatic picture of the scientist at work in the laboratory designing an experimental apparatus to "stage" a phenomenon as a "reliable" witness that authorizes the scientist to speak "in its name" (Stengers 2000, 84; 1997, 88, 139). Stengers (1997, 85) writes that for a scientist to speak authoritatively, "it is actually a matter of constituting phenomena as *actors* in the discussion, that is, not only of letting them speak, but of letting them speak in a way that all other scientists recognize as reliable." It is in this respect that "the art of the experimenter is in league with power: *the invention of the power to confer on things the power of speaking in their name*" (165).

The beauty of this argument is that it does more than correct the governmentalist fiction of Science. It prompts a rethinking of political representation as well, powerfully countering what E. E. Schattschneider ([1960] 1975, 131, 135) has criticized as the "simplistic" notion of democracy as involving popular political participation that is spontaneously generated "at the grass roots." It is incumbent on politicians, no less than scientists, to stage the forces for which they claim to speak. Their authority depends on the capacity of the political system to distinguish between measures of public opinion and popular votes that are "reliable" and those that should be disregarded as "extorted testimony" (Stengers 1997, 141). The demos, like the experimental phenomenon, is a "fact of art" (Stengers 2000, 84).[2] Consequently, the art of the politician is "in league with power" no less than that of the experimenter.

This is not the familiar power of the back room bargain or the legislative horse trade. It is what Latour (2001, 315) elsewhere calls "trials of force," a public contest that puts a politician's belief that she speaks for a following to the test.[3] In representative government as we Westerners practice it, such trials rarely occur. Political representatives are subject to the "retrospective" judgment of voters who can vote them out of office if they are displeased with their performance (Manin 1997, 179). As election is a punctual event, not an ongoing process of consultation, Western democracies allow citizens only to sanction legislators. They have no occasion to form and act on their own political judgments. Were representative democracy designed to stage "trials of force," then it would be clear, as Latour puts it, that "staging," or "the multiplication of artifices to fabricate [political] agents that can say 'yes' or 'no,'" is "at least as important a skill as the construction of facts by researchers in laboratories" (Latour 2004, 144–45).

Staging, or mediation, is where Latour and Stengers propose to connect science and politics, the laboratory and the Assembly. That connection is by no means a fallback on animism. It is a provocation to both fields that manages at once to extend the capacity for speech to things while withdrawing that of voice from humans.

By making a distinction between "speech" and "voice," I mean to join Latour in foregrounding the work of mediation that is prerequisite to speech.[4] Hannah Arendt ([1963] 1984) has called attention to this work in an oft-cited passage from *On Revolution* where she emphasizes the importance of acknowledging citizenship as artificial. She drives home this point by tracing the continuity of the modern concept of legal personhood to the Latin word *persona,* which, in its "original meaning . . . signified the mask ancient actors used to wear in a play" (106). The function of the mask was twofold: to "replace the actor's own face and countenance" and to "make it possible for the voice to *sound through*" (106). Just as the mask gives an actor a role to play and a publicly audible voice, so, too, do legally prescribed rights and duties enable a man to stand and speak before the law in ways that must be taken into account.

There is a strong affinity between Arendt's mask and Latour's notion of "speech impedimenta" *(embarrass de parole),* which he, too, introduces to discredit the fiction of spontaneous expression. That fiction, as dear to democracy as it is to phenomenology, grounds the belief in an "authentic" general will or "people's voice" as the ultimate arbiter of competing political demands (Derrida 1973). Once understood in the context of the critique of voice that underlines it, it is clear that Latour's insistence that both scientists and politicians are spokespersons is not meant to personify things but to denaturalize voice and call attention to the mediation that makes speech possible.[5] Underscoring this double move, Latour (2004, 68) writes, "We do not claim that things speak 'on their own,' since no beings, not even humans, speak on their own, but *always through something or someone else*."

This is what makes their argument so captivating: Latour and Stengers unfold a conception of representation that takes its definition from the Assembly (spokespersonship) but its practice from the laboratory (staging). The upshot for science studies is the claim that scientists represent. That for political theory is that representative democracy has something to learn from experimental science. Simply put: scientists actually do a better job of "constituting phenomena as *actors* in [scientific] discussion" than mass democracies do constituting citizens as participants in politics because, in Latour's (2004, 170) words, "we [moderns] actually know how to consult nonhumans better than humans!" Latour explains this failure by the fact that "politicians . . . imagine that one can speak of them in their place and without ever truly consulting them—that is, without ever finding the risky experimental apparatus that would allow them to define their own problems *themselves* instead of simply answering the question asked" (171). I suggest that the mythology of "voice"—the belief in spontaneous expression and spontaneous participation that Schattschneider so aptly criticized—has something to do with this failure to "truly" consult.

Modern democracy, no less than science, is plagued by ambivalence toward the practices of representation that make it possible. It comes much more easily to hold political representation to the (impossible) measure of mimesis than to acknowledge what Nadia Urbinati (2006, 119) calls its "ideological" and "rhetorical" aspects. Put in Stengers's terms, Urbinati's point is that the modern relationship between representative and constituent is "staged." This is by contrast to the organic model of feudal times when a representative was a delegate for the fixed social group to which he belonged. Urbinati counters that with the democratic revolutions of the eighteenth century, "the representative is entirely constituted by and through her political relationship with her constituency; her belonging to the constituency is an *idealized* and *artificial* construction" (118; italics added). For Urbinati, as for Stengers and Latour, to speak of artificiality in this context is not to suggest that representatives are unbound by any obligation to their constituencies but to insist that we citizens ought not to be asking whether our representatives look like us or act like us. We should be demanding a democratic system that stages "risky" interrogations between politicians and publics, Latour's "trials of force."

This is what it would mean to *faitiche-ize* the people. Recognizing that there is no authentic "voice" of the demos, we citizens of Western democracies should object when elected representatives speak in the name of that fantasy. We should be discontent when voting is the primary means of staging the popular will. Why should not modern democracy, like modern science, rest its credibility on trials of force? This would mean that the art of the politician, no less than that of the experimenter, would consist in "designing devious plots and careful staging that make an actant," in this case a citizenry or constituency, "participate in new and unexpected situations that will actively define it" (Latour 1999, 123).

Electing Scientists as Spokespersons

Latour and Stengers propose to capture this idea of mutual constitution with the term *faitiche.*[6] This is not "fetish" (although it is phonetically identical in French). That inescapably pejorative term denotes a fantastic entity, one that its followers endow with a mastery that no one can have but that they desperately desire someone or something to possess. Confronted with a fetishist, the response of the modern is to insist on a choice: "Did you make it, or is it a true god?" (cf. Latour 1996, 16). *Faitiche*, by contrast, plays on the "double etymology" of *fait,* the past participle of the French verb *faire* (to make or do), and the French noun *fait* (fact) to refuse this choice and embrace the paradox of what Latour and Stengers call constructivism: that "facts [are] facts—meaning exact—*because* they

[are] fabricated" (Latour 2005, 90).[7] The term *faitiche,* then, is a play on the ambiguity of the word *fait,* which means "in the same breath, that which someone has made and that which no one has made": "'un fait est fait,' as Gaston Bachelard put it" (Latour 1996, 38; 1999, 127).[8] Admittedly, this does not look like much of a paradox at first. It seems merely to echo the very familiar notion that "reality is socially constructed," which asserts two perfectly noncontradictory claims: first, that there is no fact uncontaminated by cultural values; second, that there exists no way of knowing or standpoint of knowledge that is not implicated in social relations of power. Social constructivism takes us straight back to the "fetish," that instrument of domination taken for a god, and the fetish, in turn, leads right into relativism: there is no authority to quell the rivalry of warring gods.

Latour and Stengers distinguish their constructivism by the insistence that it *is* possible to "talk truthfully about a state of affairs" (Latour 1999, 114). Stengers (2000, 76) has stated adamantly that she resolutely opposes the view that "no author of an abstract proposition has the means to make nature a witness in order to carry the decision concerning its truth." Humans *can* speak on the authority of nature. Human reason *does* have the "power to link up with the reason of things," although that linkage occurs neither by way of transcendent universals nor by raw experience (77). Instead, it is Galileo, as the founder of the experimental sciences, who demonstrated how that link can be activated by giving us "*something one believed to have been lost: the power to make nature speak*, that is the power of assessing the difference between 'its' reasons and those of the fictions so easily created about it" (81). Following Latour in imputing what would typically be understood as the work of politicians to the sciences, Stengers asserts that they "invent possibilities of representing, of constituting a statement that nothing a priori distinguishes from a fiction, as the legitimate representation of a phenomenon" (87).

Stengers manages at once to insist on the truthfulness of the experimental sciences and to refute realist ontology together with the positivist conception of science that accompanies it. On her account, the sciences do not research an autonomous "reality" by methods that claim objectivity because they are fully independent of the processes they "discover."[9] The sciences' relationship to nature is mediated, which is to say that it is a relationship in which each participant is understood to "*do something*" (Latour 2005, 128; cf. Stengers 2000, 99). As Ryan Holifield (2007, 103) puts it, mediators are "actors which transform as they translate." This transformation, as Stengers (2000) specifies it, involves a mutual constitution of authority. Beginning from the postpositivist premise that the authority (or facticity) of nature is "not a given," she explains that whereas "scientists recognize 'nature' as their sole 'authority,'" they know that to activate "the possibility for this 'authority' to create authority . . . it is up to them to constitute nature as an authority" (93).

Her favorite example is Galileo's inclined plane. It "represents an experimental apparatus, in the modern sense of the term, an apparatus of which Galileo is the *author*, in the strong sense of the term, because it is . . . an artificial, premeditated setup that produces 'facts of art'—artifacts in the positive sense. And the singularity of this apparatus . . . is that it *allows its author to withdraw*, to let the motion *testify* in his place" (Stengers 2000, 84). Stengers describes a passage from author of apparatus to thing back to author of apparatus. If the experiment is not sound, it will not stage an effect but merely produce or cause it. The phenomenon will not be reliable and capable of authorizing its spokesperson but rather will be "dictated by the experimental conditions" (51). If the experiment succeeds, it produces two autonomous entities.

The first of these is the phenomenon that can be counted on as a force that exists not simply in the context of this particular experimental setting but in the world. As she puts it, a successful experiment transforms "a phenomena into an 'experimental fact,' a *reliable witness*, that can discern among those who interpret it," ultimately "authorizing and supporting the thesis of the one who speaks in its name" (Stengers 2003, 56; 1997, 139). This "one who speaks in its name" is the second autonomous entity: the experimenter whom the motion authorizes to insist that other researchers have no alternative but to take it into account.[10] Hearkening back to the political idiom, we can say that a successful experiment endows a phenomenon with the capacity to elect a spokesperson to represent it in the most basic sense: to assert that it matters. For Stengers, then, it seems that representing means not speaking *for* an interest but speaking of a phenomenon *as* an interest, a proposition that must be taken into consideration by scholars in a field of study even at the risk that it ruins "years of work" (Stengers 2000, 95; 1997, 84).

This double passage—from inarticulate phenomenon to reliable witness and from experimenter to legitimate spokesperson—is what Latour wants to mark with the term *faitiche.*[11] He coins it to speak about the distinctive agency of mediation, an agency that is not localized in any particular agent but that materializes when an activity that engages actors in an exchange of properties produces something that "overtakes" them (Latour 1996, 46).[12] The orthography of the term, amalgamating "fact" *(fait)* and "fetish" *(fétiche),* calls attention to what they share: a common subterfuge that "conceals the intense work of construction that permits the truth" of each (44). Whereas "the word 'fact' seems to throw us back on external reality, the word 'fetish' to the crazy beliefs of the subject," Latour joins the two to debunk this symmetrical pretense (44). He does not reduce the one to the other but, on the contrary, underscores the truth of their shared "etymology": both facts and beliefs are made *(fait)* or worked up in practice. It is this making, working up, and working with that generates "the robust certitude" that makes it possible to act without thinking twice—which is to say, without agonizing over the lack of an independent, "transcendent" ground (44).

Although it may sound like a noun, Latour goes out of his way to define *faitiche* as a kind of movement or exchange—a passage or passing—rather than as a kind of thing.[13] He says it can be understood as "that which gives the autonomy that we don't have to beings that do not have it either but who, from the fact [of this gift], give it to us" (Latour 1996, 67).[14] Notice that autonomy has been significantly redefined from its typical sense as a quality, attribute, or property that we locate in an individual whom we credit with having achieved independence as a self-willing, self-legislating agent. As Latour (1999, 129) describes it, autonomy is not a localized capacity but a distributed agency that comes from the exchange of properties among an author, an apparatus, and a phenomenon.[15] Although Stengers has done much to illuminate this exchange, her focus on staging privileges the movement from apparatus to thing back to the author of the apparatus. As I will show, Latour elaborates Stengers's account by insisting that the laboratory trial is not sufficient to establish the scientist as a spokesperson. That passage is further mediated by writing the text of the laboratory trials and then by subjecting those trials, by the mediation of the text, to peer review. Before elaborating this notion of distributed agency, I want to ask how the concept *faitiche,* with its conceptualization of representation as mutual constitution and empowerment, would play among political scientists and political theorists.

The Problem of Responsiveness

Where representation is concerned, the touchstone work in the field continues to be Hanna Pitkin's ([1967] 1972) *Concept of Representation.*[16] There is an affinity to the work of Latour and Stengers in Pitkin's insistence that agency in representation is reciprocal. Self-evident as it may seem, this claim allowed Pitkin to make a radical intervention into the "mandate vs. independence" controversy (the question whether a political representative should be more like a delegate or more like a trustee). Pitkin recognized the need for independence on the part of the representative but recognized at the same time that the represented must be "*conceived as* capable of acting and judging for himself" (162). This is interesting for its upshot: it means that the test of good political representation cannot be its correspondence to something beyond politics, whether that be popular sentiment, will, or opinion (the delegate model), or a public good that transcends what any people may believe or say that it wants (the model of the trustee). Representing is an activity without a model, without certainty, and—in Pitkin's words—without "guarantee" (163).

Pitkin's kinship with science studies is most evident in her engagement with Edmund Burke's "Speech to the Electors of Bristol," which is a classic defense of political representation as trusteeship. Pitkin ([1967] 1972) goes after Burke not for his elitism but for his insistence that government does not represent persons but the public interest, conceived as an "unattached" abstraction. This, she astutely observes, reduces political representation to a set of epistemological problems: what are the conditions for discovering the "right answers to political problems"; how can these conditions be institutionalized politically; how to ensure a twofold correspondence between the public interest and the acts of representatives, on one hand, and the opinions of citizens, on the other (170)? Once posed in such terms, the problem of representation can be solved by precisely the sort of intervention that Latour seeks to rule out for scientists—an appeal to expertise. As Pitkin notes, Burke "sees interest very much as we today see scientific fact: it is completely independent of wishes or opinion, of whether we like it or not; it just is so" (180). Pitkin famously countered that "representation is not needed where we expect scientifically true answers, where no value commitments, no decisions, no judgment are involved. . . . We need representation precisely where we are not content to leave matters to the expert" (212). Political interests are not transcendent and "unattached"; they are the "interests *of* someone who has a right to help define them" (189). Even so, this does not give that "someone" the right to bind or instruct her representatives. Felling Burke's democratic opponents in their turn, Pitkin asserts with brisk efficiency: the "fact is that . . . in political representation, the represented have no will on most issues" (163).

So political representation is not trusteeship because there are no right answers to questions of political interest (Pitkin [1967] 1972, 189). But representatives cannot be delegates because on "most issues" citizens themselves have no answer to the question what their interests are. Would Pitkin have us give up on representative democracy altogether?

Not at all. And this is what makes her work so interesting. She maintains that even given the lack of appeal to a transcendent public good or immanent popular will, it is not unreasonable to demand that a "representative system . . . look after the public interest and be responsive to public opinion, except insofar as nonresponsiveness can be justified in terms of the public interest" (Pitkin [1967] 1972, 224). The trick is to understand that the public interest is not the "input" that creates a policy "output." Instead, it is defined in and through the process of political representation itself. Pitkin put it this way: "we assume that if the representative acts in the interest of his constituents, they *will* want what is in their interest and consequently *will* approve what the representative has done" (163; italics added). This "anticipatory" (Mansbridge 2003, 518) movement not only bears a striking affinity to *faitiche* but poses a similar problem.

If democratic self-rule is to have any meaning at all, it must be possible to acknowledge that citizen preferences and interests are made without having to concede that they are entirely made up.

Pitkin ([1967] 1972) solves this problem by attaching a condition to "responsiveness." She stipulates that in a democratic regime, "the representative must be responsive to [the represented] rather than the other way around" (140). Her concern is to ensure that "responsiveness" can only go *one* way: "the represented must be somehow logically prior" (140). Although Pitkin's conception of representation is not simply contradictory, she is at odds with herself.

She wants to remain committed to the idea of representation as an activity, in other words, to the idea that it cannot be boiled down to an epistemological question of accuracy or adequacy to something that is literally prior to its representation. To assign literal priority to the represented would reduce the representative to a mere delegate. As she recognizes, there is something generative about representing: "the national unity that gives localities an interest in the welfare of the whole is *not merely presupposed* by representation; it is also continually *re-created* by the representatives' activities" (Pitkin [1967] 1972, 218, italics added). Yet having conceded this, she is all the more anxious to ensure that representatives do not shape the world according to their own wills. Hence the uncharacteristically evasive and imprecise wording of the phrase "somehow logically prior" and the ambiguous ontological status to which the italicized phrases relegate "national unity" in the previously quoted sentence. Representatives cannot simply take national unity for granted (it is "not merely presupposed"), but they do not make it up out of whole cloth (it is "re-created" by their activities). She falls back on the *direction* of responsiveness—the "representative must be responsive to him rather than the other way around"—because it is the only way she can think of to ensure that representation does not cross what she calls the "line between leadership and manipulation" (233).

At this talk of line drawing, our champions in science studies may well be feeling nervous—and with good reason. Even though, to her credit, Pitkin ([1967] 1972) grants that this "line" is a "tenuous one, and may be difficult to draw," she nonetheless insists—as if italicizing could make it so—that there "undoubtedly *is* a difference, and this difference makes leadership compatible with representation while manipulation is not" (233). What makes her so sure? She derives her certainty from the very "basic meaning" and "correct definition" of the term *representation,* which Pitkin takes to follow from what she calls the "etymological origins" of the word: "*re-presentation*, a making present again" (8). For Pitkin and the many political scientists who have made her work the definitive source on this matter, it is as if we can see from the prefix—the *re-* in *representation*—that political representation comes after the fact, hearkening back to and "reduplicating" something that has come before (Derrida 1973, 57). To suggest that it could be otherwise, Pitkin ([1967] 1972) memorably asserts, is to court the "fascist theory of representation" whereby a leader "aligns" his followers to himself on the basis of "emotional loyalties and identification" that "need have little or nothing to do with accurate reflection of the popular will" (140, 108, 106). Notice what this cold war ideology has smuggled in: a dual-world ontology that frames the problem of political representation in terms of the (naively realist) laboratory—as a matter of adequacy to an empirical referent. Such an ontology is hostile to *faitiche.*

In a recent work that includes a short commentary on Pitkin's classic, Ernesto Laclau has proposed an approach to conceiving of political representation that our champions in science studies might find more congenial. For Laclau, as for Pitkin, political representation is an activity without "guarantee." This means to him that the activity of representing links political demands that are irreducible to a common concept or material condition. This irreducibility means that "nothing in those demands, individually considered, announces a 'manifest destiny' by which they should tend to coalesce into any kind of unity" (Laclau 2005, 162). Consequently, "they do not tend *spontaneously*" to come together but must be brought together by "adding something" to the represented, an "addition [that], in turn, is reflected in the identity of those represented" (108, 158; italics added). Because Laclau agrees with Pitkin that representation cannot simply presuppose the "unities" for which they stand; he must disagree with her that responsiveness can only go one way. He counters that there is an "essential impurity in the process of representation" because the process of representation is necessarily "two-way": it is "a movement from represented to representative, and a correlative one from representative to represented" (Laclau 1999, 98; 2005, 158).

By his emphasis on movement, Laclau comes much closer to embracing *faitiche*. Yet his account of movement is exclusively discursive and abstract. He theorizes this "two-way" process as "naming," proposing that heterogeneous demands are brought together into a relatively unified collective agency by virtue of the "social productivity of a name" (Laclau 2005, 107, 108). A name is *faitiche*-like in that it is not simply arbitrary (it must succeed as a hail or, in other words, be taken up if it is to have any purchase politically), but neither is it organic. It does not correspond to a common essence or need that the various parties to an alliance actually share despite their apparent heterogeneity. By contrast to Pitkin, Laclau holds nothing to be prior to this act of representation—neither literally nor "somehow logically." On the contrary, he contends that "names retrospectively constitute the unity of the object" (Laclau 2005, 163). Names are subject to trials of force. As Oliver Marchart (2005, 9) has put it, they bring agencies into being by a "hegemonic intervention."

Has Laclau not just proven Pitkin's point (and illuminated the danger of *faitiche*)? Has he not demonstrated that breaking the condition of unidirectionality courts fascism? To those who hold what John Mowitt (2002, 51) calls a "romantic" conflation of agency with the "agent" (the idea of a deliberative consciousness "fundamentally grounded in a will that can assert itself against whatever contextual forces might be said otherwise to constrain it"), the answer would have to be yes. Acknowledging the "impurity" of representation poses the problem that the agency of the represented—"except at the level of party militancy, with all that it entails: science, organic intellectuals, discipline, etc—is all but foreclosed" (49). What is produced by naming "is wholly a *subject* (decidedly not an agent)" and so "harbors the ultimate political danger" to democratic politics (51; italics added). But what if one does not hold to such a "romantic" ideal? Mowitt's distinction between agency and agent helps me to formulate what I believe Pitkin sought but could not say, and what Laclau put words to but whose practice he did not theorize. This is what Latour and Stengers explore through the concept of *faitiche* and in the practice of the laboratory: how to talk about agency that is not seated in a subject but rather distributed throughout a system of representation or a field of action.

Writing Agency

Latour's account of distributed agency puts three modes of representation in play: political, symbolic, and juridical. He contends that before a scientist can represent (or speak for) the phenomenon he has staged, she must first represent ("narrativize") the experimental situation and then advocate for and defend the experiment to her peers.[17] Latour emphasizes that in any experiment, two different kinds of questions are at stake. There is the "ontological" question whether the experiment has "conveyed" anything new "to modify what [a scientist's] colleagues say about him and about the abilities of living organisms that make up the world" (Latour 1999, 123).[18] There is also the question of the "epistemological" status of the statements to which it gives rise: is the entity a "reliable witness" to, and the scientist a legitimate "spokesperson" for, a phenomenon, or was it just "an amusing story" (123)?

To keep all these aspects in play, Latour (1999, 123) defines an experiment as a "*movement*" of three distinct trials. Not all these trials take place in a laboratory. Moreover, as they involve both "literary" and "nonverbal, nonlinguistic" but emphatically "artificial" components, there is more to this than what a positivist would count as experimentation (123–24). Latour calls the first trial "a story" of transformation that is "similar to any trial in fairy tales or myths" (123). The second is "a situation" composed of the apparatuses the scientist uses to design "devious plots" to isolate the properties of the entity and stage it to "define" itself by modifying, transforming, perturbing, or creating other beings (123). The measure of a successful experiment turns on the extent to which the story and situation can be distinguished. Can the first, which possesses avowedly "literary aspects," be shown not to have contaminated the second, which should pertain exclusively "to nonverbal, nonlinguistic components" (123)? This question constitutes the third trial, the trial of peers, in which the scientist must "convince the Academicians that [the] story is not a story," in other words, not just a story, that the "competence" of the new entity "is *its* competence, *in no way* dependent on his cleverness in inventing a trial that allows it to reveal itself" (123). If the experiment "withstands the Academy's scrutiny, then the text itself will be in the end authorized by the [entity], the real behavior of which can then be said to *underwrite* the entire text" (132). The situation (the experimental trials) detailed by the scientist will be taken to warrant the story he has told about the phenomena.

Latour (1999) sums this up in a formulation that might well perturb both a constructivist and a realist: "an experiment is a text about a nontextual situation, later tested by others to decide whether or not it is simply a text. If the final trial is successful, then *it* is not just a text, there is indeed a real situation behind it, and both the actor and its authors are endowed with a new competence" (124). An experiment is a "text" (here the realist shudders) about a "nontextual situation" (now the constructivist will twitch) tested to decide whether it is "simply" a text or whether it has "a real situation *behind* it" (sigh). Which is it? Ontological dualism? Radical constructivism? I think neither.

Put differently, but in ways that Latour's own terminology here justifies, a successful experiment conveys a new fact that has the status of an indexical sign. An indexical sign—the movement of the needle on a barometer, for example, or the mercury in a thermometer—can be said to carry the "force of the real" in the sense that its meaning cannot be determined exclusively in relation to other signs.[19] It is, rather, the effect of an extralinguistic force that requires some kind of artifice or staging to manifest itself. Whereas the meaning of a Saussurean sign is entirely conventional, which is to say entirely internal to a linguistic system, in the case of the indexical sign, a device or artifice not only stages the action of something outside that system but typically also transposes it into a linguistic system. In the case of the thermometer, mercury indexes temperature and enables it to be spoken about in terms of degrees. What must be underscored is that the apparatus that are so crucial to staging are at once "nonverbal, nonlinguistic" and signifying. In other words, the devices whereby facts and indexical signs come to be are not just physical apparatus like Galileo's inclined plane; they are also (and often at the same time) signifying systems. And then there is one further step: narrativizing what went on. This suggests, as I contend that Latour's own reading of Pasteur's "Mémoire sur la fermentation appelée lactique" shows, that the devices whereby facts come to be are experimental, rhetorical, and literary all at once.

Latour reads to Pasteur's "Memo" not as the canonical text that historians of science take it to be but as an autoethnography of the laboratory. What makes it so useful to him is that it dramatizes the distribution of agency that Latour attempts to capture with the term *faitiche.* Latour contends that Pasteur establishes the "experiment as an event" in which "the actor and its authors have been endowed with a new competence: Pasteur has proved that the ferment is a living thing; the ferment is able to trigger a specific fermentation different from that of brewer's yeast" (Latour 1999, 124). Pasteur has "given" an autonomy to the ferment that he comes to possess himself only by virtue of being authorized by the ferment whose movement he staged. In effect, Pasteur's "Memo" sums up "the mystery of the two opposed meanings of the little word 'fact'" (115, 127, 129). That is not to say that it establishes the ferment as "at once *fabricated* and *real,*" an assertion that is at once utterly banal and absolutely without drama because it could be said of any industrially or hand-crafted object (such as a pencil or a chair). Instead, it dramatizes that by virtue of having been "artificially made up" by Pasteur in his laboratory, the lactic acid ferment gains "a complete autonomy from any sort of production, construction, or fabrication" (127). This autonomy is the "miracle" to be explained, how it is that the "obvious immanence" of a laboratory-made fact does not "run counter to its validity and truth" but are the source of "its downright transcendence" (129).

There is a sense in which Pasteur's problem and that which I derived from Laclau are the same. From an amorphous catalog of heterogeneous effects, Pasteur moves to a name: "lactic acid ferment." A positivist might insist that this entity was there all along, that Pasteur has merely provided a new way of looking at fermentation and attached a new label to it. Latour (1999) would counter that, to the contrary, Pasteur has ruled out the old way of speaking about fermentation (as decay) and constrains all those who come after him to accept that the ferment is not an inert "chemical mechanism" but a vital "entity in its own right" (116). In Ryan Holifield's (2007) terms, Pasteur has not simply offered a "new interpretation" of fermentation; he has "materially reassembled" the ferment. This goes beyond "naming" to a material, practical assemblage of forces that Stengers would call "reliable." Speaking politically, we would say that such a force can be "called" by name and can be trusted to say yes or no on its own terms.

If a plausible account of such a miracle is to be had, it is not to be found by reading philosophy but by "div[ing] even deeper into some empirical sites to see" what scientists actually do to "get out of the difficulty" (Latour 1999, 127). This is what Pasteur's "Memo" allows Latour to do. Read as an autoethnography, it "beautifully illustrates" what Latour calls the "very simple setup" that lets Pasteur move between a fact as "experimentally made-up" and a fact as "*not* manmade" (125). Latour observes that Pasteur does this without giving it a second thought, putting forward in a single paragraph what must be taken as "*entirely unrelated epistemologies*" (129). Pasteur juxtaposes a "confession of . . . prejudice"—that "facts need a theory if they are to be made visible, and this theory is rooted in the previous history of the research program"—against an assertion of independent, context-transcending validity: that the "results" of his research can be judged "impartially" to be true (129). Latour finds in the "Memo" the answer to the question, how does Pasteur cross the "gap between [these] two opposing sentences" (128)?

Latour (1999) emphasizes that Pasteur neither "eras[es] the traces of his own work as he goes along" nor takes his own vigorous activities ("extracting, treating, filtering, dissolving, adding, sprinkling, raising the temperature, introducing carbonic acid, fitting tubes, and so on") to diminish the autonomous capacities of the yeast (132, 131). With Pasteur first narrating himself in action and then narrating the action of the ferment as if the two supported rather than competed against one another, Latour asserts that the "experiment creates two planes: one in which the narrator is active, and a second in which the action is delegated to another character, a nonhuman one" (129). It then "*shifts out* action from one frame of reference to another" so that agency is distributed across both (129).[20] It may be tempting to pose to Pasteur the question by which fetishes are smashed: "Did you make it or is it real?" But he will refuse this choice. Or Latour will refuse it on his behalf: "Pasteur acts *so that* the yeast acts alone" (129).

Latour (1999) seems to suggest that if Pasteur himself has no trouble moving back and forth between foregrounding his own activity and foregrounding that of the yeast, between offering a realist account of his actions and a constructivist one, then neither should a philosopher of science.[21] But there is a doubling of Pasteur that Latour loses sight of. There is Pasteur the scientist who works with the devices we would expect to find in a laboratory—"he sprinkles, boils, filters, and sees"—and there is Pasteur the "narrator" whose active employment of literary and rhetorical devices Latour recognizes but never enumerates. On the contrary, in his analysis of the crucial passage in which Pasteur describes the "main experiment," Latour occludes those devices.

Latour's (1999) reading of this paragraph emphasizes, in keeping with the *faitiche* concept, how Pasteur spotlights his own activities, convinced that this will not compromise "the autonomy of the entity 'made up' inside the laboratory walls" (127). Nowhere is this more evident than "at the very moment where the entity is at its weakest ontological status" and "the experimental chemist . . . *in full activity*" (131). Latour quotes Pasteur, italicizing the details and pronouns that Pasteur uses to mark his actions: "I *extract* the soluble part from the brewer's yeast, by *treating* the yeast for some time. . . . The liquid . . . *is carefully* filtered . . . one *raises* the temperature. . . . It is also *good* to *introduce* a current of carbonic acid" (131). Latour sums up this account of the "experimental chemist" at work in the lab, "extracting, treating, filtering, dissolving, adding, sprinkling, raising the temperature, introducing carbonic acid, fitting tubes, and so on" (131). The paragraph ends with a "shifting out," a change of time, and a crucial move from *I* to *we* that changes the frame of reference from the activities of Pasteur to those of the ferment: "On the very next day a lively and regular fermentation is manifest. . . . In a word, we have before our eyes a clearly characterized lactic fermentation" (131; quoting Pasteur). This is Latour's comment: "the director withdraws from the scene, and the reader, merging her eyes with those of the stage manager, *sees* a fermentation that takes form at center stage *independently* of any work or construction" (131–32).

Really? Is there no "work or construction" in the shift from *I* to *we?* Does "the reader" merge "her eyes," or does the narrator change the subject? Is there no artifice in the transitional device "on the very next day"? Is there no rhetorical art to "withdrawal"? Did Pasteur not attempt to absent himself earlier with his shifts from speaking in the first person as "I," to the third person "one," and finally to the abstract impersonal pronouncement "it is also good"? Certainly there is construction involved in the very stinting of detail, as Pasteur offers "stylized accounts" of experiments that are not at stake but reintroduces "human agency" with a "recipe-like description" of the procedure for lactic acid fermentation on which he relies as the "stabilized" procedure from which his living yeast "will be made to appear" (Latour 1999, 131).[22]

What is striking about Latour's rendering of this section of Pasteur's "Memo" is that although he flags Pasteur's rhetorical devices at every turn, the only activities Latour describes him engaged in are those associated with experimental chemistry. Latour recognizes Pasteur as author. He takes note of Pasteur's use of narrative devices to lend depth to the text, to create within it an effect of crosscutting from one site of action to another and from one day to the next. He emphasizes that language does work in the "Memo" both by employing the concept of "shifting out" and by using the term *experimental scenography,* which he coins to foreground the effect of the shifts (Latour 1999, 129). Yet from the lists of activities in which he describes Pasteur as engaged, the author–narrator disappears. The pertinent agencies are that of the scientist as experimenter and that of the ferment as living entity. The scientist as narrator, with his literary devices, leaves traces everywhere, but Latour does not mark them as activities.

Latour (1999, 127) insists that it is in the "empirical site" of the laboratory that the mystery of *faitiche*, the handoff between realism and constructivism, the passage between facts that are contingent on theories and facts that arbitrate among them, can be solved. His account suggests otherwise. Pasteur smoothes this passage not just by what he does but by the way he composes it into a story. He uses "literary technology" to compel the assent of a "virtual" witness (Shapin 1984, 491).

Shapin (1984, 483) has proposed "virtual witnessing" to account for how experimental scientists "produced the conditions" for mobilizing universal assent to "matters of fact" with the advent of the "probabilistic conception of knowledge" in the mid-seventeenth century. If knowledge were probable, not certain, it could not command assent like a geometric proof. Henceforward, it would be validated by being witnessed by a multiplicity of spectators. Shapin emphasizes that "literary technology," the use of expository conventions to produce "in a reader's mind . . . such an image of an experimental scene as obviates the necessity for either its direct witness or its replication," became crucial to the constitution of virtual witnesses (491). Literary technologies recruited publics who would have occasion neither to see an experiment performed nor to replicate it in a laboratory. Such technologies include the use of illustrations and a prose style designed to imitate mimetic representation. Complex sentences packed with circumstantial detail "were to be taken as undistorted mirrors of complex experimental performances" (494). The goal of such an account was not just to produce the effect of a text that mirrors the experiment but also to secure the more crucial illusion of the experimental phenomenon itself as "the very mirror of nature" (507).

Seventeenth-century rhetorical performances were opposed to *faitiche,* being designed to create the "illusion that matters of fact are not man-made" (Shapin 1984, 510).[23] As Latour reads the "Memo," Pasteur went out of his way to resist these seventeenth-century conventions. Why does Latour not more explicitly thematize the literary technologies he employed to do so?

To show why this question is significant, let me rehearse once more the distinction Latour (1999) proposes to characterize the measure of a successful experiment. He contends that the scientist must convince the Academy that the avowedly "literary" transformation "story" of the experiment did not contaminate the "nonverbal, nonlinguistic" components of its "situation" (123, 124). In the problematic paragraph I quoted earlier as perturbing both the constructivist and the realist, Latour presents this distinction between "story" and "situation" as if it were given. Perhaps it is grounded on the representational divide that Latour creates in separating the "literary" from the "nonlinguistic." I have already noted, in suggesting that an experimental phenomenon be conceived as an indexical sign, that such a separation is untenable. An experimental apparatus simultaneously stages an action (i.e., displays a phenomenon that exists independently of the apparatus) and transposes it into a system of signification (even one as rudimentary as a thermometer). In light of this simultaneity, the difference that Latour asserts between the terms "story" and "situation" takes on a different cast. Far from marking the distinction between the "literary" and "nonlinguistic" elements of the experiment, it is itself an effect of Pasteur's narrativization of the experiment as a story that "tells itself," a staging that unfolds before the eyes of a public that Pasteur's literary technology recruits as a virtual witness (Chambers 1984, 32).[24]

Latour claims that the "mystery" of *faitiche* can be answered by the distribution of agency across the two planes that the "experiment creates." More precisely, it is Pasteur's "literary technology" that creates these planes. By stinting detail, changing the subject, and occupying different positions of enunciation, Pasteur manages alternately to foreground his own activity and that of the ferment. Pasteur's authority as spokesperson depends not only on the agency of the ferment, as mediated by the passage of the experiment that he has made, but also on the representation of the experiment, whose passage into narrative he has also made. Although Latour certainly does not deny the rhetorical and literary aspects of experimental science, he does not explicitly mark them as activities in which Pasteur is engaged. Perhaps to do so would be to compromise the distinction between story and situation—a distinction to which he is as committed as Pitkin is to her distinction between leadership and manipulation. Perhaps Latour has something to teach Pitkin that he has not fully accepted himself: whether science or politics is at issue, that there is no stripping representing of its literary devices.[25]

Faitiche-izing the People

Is representation *faitiche*-ization? The sound of the word suggests a purely symbolic politics whereby a charismatic leader puts responsiveness in reverse, remaking a people in his own image. Yet if we follow Latour's argument about mediation, it suggests that the trouble with fascism does not have as much to do with the direction of responsiveness as Pitkin imagines. It has rather to do with the fantasy of an *unmediated* relationship between represented and representative that fascist regimes render plausible by orchestrating consensus in a variety of settings. To give just a couple of examples, there is the abolition of the juridical apparatus that institutionalizes dissent (such civil liberties as freedom of speech, press, and assembly plus a robust legal conception of privacy), and its replacement with a surveillance apparatus that knows no bounds. Politically, there is the dismantling of a partisan apparatus that produces legitimacy by fostering competition for power (pluralism of political parties and interest groups, regular contested elections) and its replacement by public shows of unanimity: mass rallies where a people shouts and moves in unison when cued by its leaders.

Responsiveness can go two ways so long as it is mediated in the specific sense that Latour (2005) gives to that term. Mediation denotes a passage that sets up a risky "concatenation" of actors, each one of which possesses the capacity to make the others "do *other* things than was expected" (59). Latour's notion of mediation has the potential to make a powerful intervention into contemporary efforts to reform representative democracy. These efforts, most notably in France, which recently enacted a law requiring gender parity on party lists, have tended to focus on what a legislature should look like to be truly representative of its constituency. The European Union has seen a resurgence of support for descriptive representation, the idea that social groups (whose composition is taken as given) should be served by representatives who resemble them. Such demands rest implicitly on an epistemological conception of representation as involving accurate correspondence to a referent. The work of Stengers and Latour lends itself to a dynamic conception of representation and a contestatory understanding of democracy. They would not be content with a legislature that matches the demography of its constituencies, nor with the simple fact of congruence between a legislator's votes and the policy preferences of his district.[26] Representative democracy would *faitiche*-ize the people in the sense of distributing agency on both sides of the legislator–constituent relationship. The U.S. political system is so far from such an arrangement that the required changes are as obvious as they are unlikely. To begin with, consider uprooting the electoral protections that secure the two-party system against risk and reforming procedures of legislative districting that divvy up safe seats among the two major parties. It would certainly also be important to overhaul the campaign finance and election system, to trust-bust the corporate media, and to tighten restrictions on campaign advertising (or to outlaw it altogether).[27] It should be as possible for actants in politics as it is for those in science to have a dialogic relationship with those who claim to represent them, one where legitimacy is based on openness to risk rather than on its foreclosure.

Notes

1 Whereas Stengers affirms the notion of scientist as representative, she asserts a clear distinction between politics and scientific activity. She contends that science, "as opposed to other activities that bring people together," has "the means of coming to relative agreement" by virtue of the fact that the influence of a scientist depends on his acting as a representative for a "third party," which is the experimental phenomenon (Stengers 1997, 88).

2 This is not to say that they are made *up* but simply that they are *made*. Stengers (2000, 99) writes, "All experimental facts are 'artifacts,' but because of this they give meaning to the tests whose vocation it is to *assess the difference between artifacts*—tests that disqualify artifacts that are said to be purely relative to the protocol that created them, and accept artifacts that are said to be 'purified' or 'staged' by this protocol," and are durable enough to "be put to the test by other questions."

3 My translation. Latour's (2001, 315) French is "*épreuves de force*." The English translation has "trials of *strength*," which, though not an incorrect way to translate force, muddies the distinction that Latour (1988, 210) attempts to create between force, which depends on allies, and potency, which is an illusory property of an individual.

4 This distinction is only implicit in his work. Stengers sometimes gestures toward it by putting "saying" in scare quotes but does not explain why she does so (e.g., Stengers 2000, 10).

5 Schaffer (1991, 182) oversimplifies in accusing Latour of "hylozoism, an attribution of purpose, will and life to inanimate matter, and of human interests to the nonhuman."

6 Latour (1996) presents this idea in a short monograph, *Petite Reflexion sur le Culte Moderne des Dieux Faitiches.* This imprint—delightfully named "those who prevent thinking in circles"—seems to specialize in publishing works by academics that appeal to a broader public audience (they have also published several works by Stengers, including *Cosmopolitiques*). Although parts of this work appear in *Pandora's Hope,* neither the translation nor the edit do the original piece justice; consequently, I will cite the French version. All translations are my own.

7 Ian Hacking (1983, 167) has underscored this point by emphasizing the amount of making that goes into the "observation" of a fact. Hacking notes that theories of science—under the influence of positivism—have both misunderstood observation by conceiving it as "reporting-what-one-sees" and exaggerated its importance in the practice of laboratory science: "Often the experimental task, and the test of ingenuity or even greatness, is less to observe and report, than to get some bit of equipment to *exhibit* phenomena in a reliable way" (170; italics added). Far from being a spontaneous and unmediated relationship with a "real" world, observation as experimental scientists understand it involves "staging" by instruments and apparatus. It follows that "observability does not provide a good way to sort the objects of science into real and unreal" (170). Even the strictest separation between theory and observation (were it possible, as positivism insists it must be) would not guarantee access to phenomena in themselves because so many phenomena in today's experimental sciences can only be observed if they are made visible by the elaborate stagings of expensive laboratory apparatus.

8 This double etymology is not difficult to convey to someone who has had introductory-level French. The term *fait* can be the past participle or the third person singular of the verb *faire* (i.e., "made," "did," "makes," or "does"), which, in French as in English, is used not only to talk about what I make or do but what I have someone or something else do *(fait faire).* The full text in *Pandora's Hope* actually makes this confusing by translating *fait* as "fact" so that the sentence reads "the double meaning of the word 'fact'—that which is made and that which is not made up" (127). It would be difficult for a reader who was unfamiliar with basic French to understand how the word *fact* bore this double meaning *in itself,* that is,

independently of the premise that it is being invoked to substantiate (that facts are facts *because* they are made).

9 It follows, as she elaborates elsewhere, that "facts" cannot be conceived as "a common material whose ideal vocation would be to assure the possibility of a comparison or confrontation" among rival hypotheses as in the "logicist or normative mise-en-scène" (Stengers 2000, 50).

10 There are similarities between Stengers's notion of an "experimental fact" and Hacking's (1983, 262) discussion of "experimental entities." Hacking contends that what makes a phenomenon real is not that it is known for what it *is* but for what it *does*. Hacking presents it as a commonplace of experimental science that "entities that in principle cannot be 'observed' are regularly manipulated to produce a new phenomena *[sic]* and to investigate other aspects of nature. They are tools, instruments not for *thinking* but for *doing*" (262; italics added). They qualify as instruments for doing because experimentation has proven them to be reliable, and scientists have come to accept that they must take them into account. Hacking observes that although nuclear physicists cannot say what an electron looks like, they know what it acts like. They have "come to understand some of the causal powers of electrons" to the point where it is possible to "build devices" that use electrons to stage "well-understood effects in other parts of nature" (262). It is this capacity to use the electron to produce reliable effects that makes it impossible to disbelieve in its existence: the electron "has ceased to be theoretical and has become experimental" (262).

11 For Stengers (2003, 39), the term *faitiche* captures a "paradoxical mode of existence," exemplified by the neutrino, "of all of those beings that, at once, have been constructed by physics and exist in a mode that affirms their independence in relation to the time of human knowledge." As she interprets the term, it could stand in for her "reliable witness" or "experimental fact." By contrast, Latour glosses it as a term for certain "types of action" rather than as a mode of being.

12 Pursuing the sport metaphor, think of how important passing is and what a (rare) skill it is for a player to know when and where to pass. Passing the ball permits an exchange of properties among players who are endowed with various distinct skills: defense, moving the ball, stealing it, scoring from far out, scoring at close range. The agency resides neither in the ball nor in the players but is distributed

among various mediators: the players, the ball, the rules of the game, the surface on which it is played, the coaches, and the umpires (to name just a few).

13 Its English translation as "factish" makes it sound like an adjective.

14 He calls it "*la sagesse de la passe*," an idiosyncratic expression that is a bit difficult to translate because the verb *passer* has many meanings in French (as it does in English). Literally, it would most likely translate as "wisdom of the pass," possibly also of "passing" or possibly "getting by."

15 I read the experimental setting as a comedy of manners. The author of the apparatus defers to the motion it stages; the phenomenon defers to the apparatus that solicits its testimony; the apparatus defers to the experimenter who authored it. All along there are little blunders—mistaken identity, the slamming of doors. And when the whole story is written up—if Latour does the writing—it is done in such a way as to emphasize the miscommunications. Autonomy is this shy, awkward, and even apologetic passage from hand to thing to hand to text, a comedy that is clarified by the experimental setting but not unique to it. Open a door, drive a car, write a sentence, cook a meal: with every competence there will be an exchange of properties between hands and things and a story that either erases or accentuates the points of resistance.

16 Nadia Urbinati's (2006) *Representative Democracy: Principles and Genealogy* is the first text in over forty years to challenge Pitkin's conception of representation. An original and provocative work, Urbinati's dynamic and rhetorical model of representing has important affinities with science studies whose elaboration are beyond the scope of this essay, although I have tried to indicate them.

17 I am using Hayden White's (1987) term *narrativize* here because I think it best captures the way Latour sees Pasteur's "Memo." It is not simply a recitation or report of facts but a transformation story whose elements Latour characterizes as "brave and surprising" and "defiant" (116–17).

18 Latour uses *him* here because he is referring explicitly to Pasteur, not because he is in the habit of assuming that all scientists are male or of using male pronouns to signify persons in general.

19 I am indebted to Andreas Gailus for this insight about the distinctiveness of the indexical sign and the phrasing of these sentences.

20 Latour (1999, 310) provides a definition for this crucial concept

shifting out in his glossary as a term "from semiotics to designate the act of signification through which a text relates different frames of reference (here, now, I) to one another."

21 Latour (1999, 131) observes, "In a single scientific paper the author may go through several philosophies of experiment with relativist or constructivist moments preceded by brutal denials of the role of instruments and human interventions and followed by positivist declarations."

22 This is a variation on Barthes's (1989) "reality effect" with two differences. First, Pasteur aims not principally to authenticate the experiment as 'real,' a direct referent of words that do no more than describe it, but to mark a distinction between the procedures he invents and those he can take for granted as well known and established. Second, he achieves this aim by a stinting of detail rather than by its superfluity.

23 Shapin (1984) does not sign on to this illusion but emphasizes that neither the experiment nor its representation were literally mimetic. It is "vital to keep in mind that the contingencies proffered in Boyle's circumstantial accounts represent a *selection*" (494).

24 Chambers's (1984) *Story and Situation* is an illuminating discussion of how realist narrative constitutes this distinction.

25 Bloor (1999) remarks on a different but related reluctance by Latour to acknowledge discursive mediation in the fact that he "makes no systematic distinction between nature and beliefs about, or accounts of, nature. He repeatedly casts the argument, his own as well as that of his opponents, in terms of nature itself rather than beliefs about it" (87).

26 Since the 1960s, political scientists have taken "congruence" to be an indicator of government's responsiveness to the citizenry; see Miller and Stokes (1963).

27 The U.S. Supreme Court in its first (2006–7) session under the direction of Chief Justice Roberts has just moved in the opposite direction, reversing an earlier ruling on the McCain-Feingold campaign finance law to fortify the First Amendment protections accorded to corporate- and union-sponsored campaign ads as core political speech.

References

Arendt, Hannah. [1963] 1984. *On Revolution*. New York: Penguin.

Barthes, Roland. 1989. *The Rustle of Language*. Trans. Richard Howard. Berkeley: University of California Press.

Bloor, David. 1999. "Anti-Latour." *Studies in the History and Philosophy of Science* 30, no. 1: 81–112.

Braun, Bruce. 2002. *The Intemperate Rainforest*. Minneapolis: University of Minnesota Press.

Chambers, Ross. 1984. *Story and Situation*. Minneapolis: University of Minnesota Press.

Derrida, Jacques. 1973. *Speech and Phenomena, and Other Essays on Husserl's Theory of Signs*. Trans. David B. Allison. Evanston, IL: Northwestern University Press.

Hacking, Ian. 1983. *Representing and Intervening: Introductory Topics in the Philosophy of Natural Science*. Cambridge: Cambridge University Press.

Holifield, Ryan. 2007. "Spaces of Risk, Spaces of Difference: Environmental Justice and Science in Indian Country." PhD diss., University of Minnesota.

Laclau, Ernesto. 1999. *Emancipation(s)*. New York: Verso.

———. 2005. *On Populist Reason*. New York: Verso.

Latour, Bruno. 1988. *The Pasteurization of France*. Trans. Alan Sheridan and John Law. Cambridge, Mass.: Harvard University Press.

———. 1996. *Petite Reflexion sur le Culte Moderne des Dieux Faitiches*. Paris: Les Empecheurs de Penser en Rond.

———. 1999. *Pandora's Hope*. Cambridge, Mass.: Harvard University Press.

———. 2001. *Pasteur: Guerre et Paix suivi de Irréductions*. Paris: La Découverte.

———. 2004. *Politics of Nature: How to Bring the Sciences into Democracy*. Trans. Catherine Porter. Cambridge, Mass.: Harvard University Press.

———. 2005. *Reassembling the Social: An Introduction to Actor–Network Theory*. Oxford: Oxford University Press.

Manin, Bernard. 1997. *The Principles of Representative Government*. Cambridge: Cambridge University Press.

Mansbridge, Jane J. 2003. "Rethinking Representation." *American Political Science Review* 97, no. 4: 515–28.

Marchart, Oliver. 2005. "In the Name of the People: Populist Reason and the Subject of the Political." *Diacritics* 35, no. 3: 3–19.

Miller, Warren E., and Donald E. Stokes. 1963. "Constituency Influence in Congress." *American Political Science Review* 57, no. 1: 45–56.

Mowitt, John. 2002. *Percussion: Drumming, Beating, Striking.* Durham, N.C.: Duke University Press.

Pitkin, Hanna. [1967] 1972. *The Concept of Representation.* Berkeley: University of California Press.

Schattschneider, Elmer E. [1960] 1975. *The Semisovereign People.* Fort Worth, Tex.: Harcourt Brace.

Schaffer, Simon. 1991. "The Eighteenth Brumaire of Bruno Latour." *Studies in the History and Philosophy of Science* 22, no. 1: 174–92.

Shapin, Steven. 1984. "Pump and Circumstance: Robert Boyle's Literary Technology." *Social Studies of Science* 14: 481–520.

Stengers, Isabelle. 1997. *Power and Invention.* Trans. Paul Bains. Minneapolis: University of Minnesota Press.

———. 2000. *The Invention of Modern Science.* Trans. Daniel W. Smith. Minneapolis: University of Minnesota Press.

———. 2003. *Cosmopolitiques I.* Paris: La Découverte.

Urbinati, Nadia. 2006. *Representative Democracy: Principles and Genealogy.* Chicago: University of Chicago Press.

White, Hayden. 1987. *The Content of the Form.* Baltimore: Johns Hopkins University Press.

Contributors

Andrew Barry is reader in geography and a fellow of St. Catherine's College in the University of Oxford. He is author of *Political Machines: Governing a Technological Society* (2001) and a coeditor of *The Technological Economy* (2002) and *Foucault and Political Reason: Liberalism, Neo-Liberalism, and Rationalities of Government* (1996). His recent research has been on the political geography of oil.

Jane Bennett is professor of political science at Johns Hopkins University. She is a founding member of the journal *Theory and Event* and the author of *Unthinking Faith and Enlightenment* (1987), *Thoreau's Nature* (1994), *The Enchantment of Modern Life* (2001), and most recently *Vibrant Matter: A Political Ecology of Things* (2010). She is currently pursuing a study of Walt Whitman's materialism.

Bruce Braun teaches social theory and political geography at the University of Minnesota. He is the author of *The Intemperate Rainforest: Nature, Culture, and Power on Canada's West Coast* (Minnesota, 2002) and the coeditor of *Remaking Reality: Nature at the Millennium* (1998) and *Social Nature: Theory, Practice, Politics* (2001). He is currently working on the urbanization of nature and on the politics of biosecurity.

Stephen J. Collier is assistant professor in the Program in International Affairs at the New School. He is the author of *Post-Soviet Social: Neoliberalism, Social Modernity, Biopolitics* (forthcoming) and has coedited *Biosecurity Interventions* (2008) and *Global Assemblages* (2004).

William E. Connolly teaches political theory at the Johns Hopkins University, where he is Krieger-Eisenhower Professor. His recent publications include *Neuropolitics: Thinking, Culture, Speed* (Minnesota, 2002), *Pluralism* (2005), and *Capitalism and Christianity, American Style* (2008). An interview in which he discusses his recent work and future projects is available in David Campbell and Morton Schoolman, eds., *The New Pluralism: William Connolly and the Global Condition* (2008).

Rosalyn Diprose is professor of philosophy at the University of New South Wales, Sydney, Australia. She is the author of *Corporeal Generosity: On Giving with Nietzsche, Merleau-Ponty, and Levinas* (2002) and coeditor (with Jack Reynolds) of *Merleau-Ponty: Key Concepts* (2008). Her chapter in this volume is part of a research endeavor on "biopolitics and phenomenology," which includes papers published in *Philosophy and Social Criticism*, *Hypatia*, and *Security Dialogue*.

Lisa Disch is professor of political science and women's studies at the University of Michigan, where she teaches contemporary continental and democratic theory as well as feminist thought. She is the author of *Hannah Arendt and the Limits of Philosophy* (1994) and *The Tyranny of the Two-Party System* (2002).

Gay Hawkins is a research professor in the Centre for Critical and Cultural Studies at the University of Queensland, Australia. Her most recent books are *The Ethics of Waste: How We Relate to Rubbish* (2006) and, coedited with Stephen Muecke, *Culture and Waste: The Creation and Destruction of Value* (2002). She is currently working on an international study of the social and material life of bottled water titled *Plastic Water*.

Andrew Lakoff is associate professor of anthropology, sociology, and communication at the University of Southern California. He is the author of *Pharmaceutical Reason: Knowledge and Value in Global Psychiatry* (2006) and editor of *Disaster and the Politics of Intervention* (2010).

Noortje Marres is a research fellow in science and technology studies at the Institute for Science, Innovation, and Society, University of Oxford. She was trained in the sociology and philosophy of science and technology at the University of Amsterdam and conducted part of her doctoral research at the École des Mines, Paris, for a thesis about issue-centered concepts of public participation in technological societies. She is currently completing a monograph titled "Engaging Devices: Participation after the Object Turn."

Isabelle Stengers teaches philosophy at the Université Libre de Bruxelles. Her interests center on the constructive adventure of the modern sciences and the challenge of embedding diverging knowledge practices in a democratic and demanding environment. She has developed her taste for a speculative, adventurous constructivism in relation with the philosophy of Gilles Deleuze, Alfred North Whitehead, and William James and the anthropology of Bruno Latour. She writes in French, but *Order out of Chaos* (with Ilya Prigogine, 1984), *A Critique of Psychoanalytical Reason* (with Léon Chertok, 1992), *A History of Chemistry* (with Bernadette Bensaude-Vincent, 1996), *Power and Invention: Situating Science* (Minnesota, 1997), and *The Invention of Modern Science* (Minnesota, 2000) have all appeared in English translation.

Nigel Thrift is professor and the vice chancellor at the University of Warwick, visiting professor at the University of Oxford, and emeritus professor at the University of Bristol. His main research interests are in the study of cities, spatial politics, nonrepresentational theory, and the history of time. His recent books include *Knowing Capitalism* (2005), *Non-Representational Theory* (2007), and *Shaping the Day* (2009).

Sarah J. Whatmore is professor of environment and public policy at the University of Oxford. She has published widely on the intersections of science, law, and democracy in relation to the contestation of environmental knowledge and rights. Her books include *Hybrid Geographies* (2002) and *Using Social Theory* (2004).

Index